IFCoLog Journal of Logics and their Applications

Volume 4, Number 7

August 2017

Disclaimer

Statements of fact and opinion in the articles in IfCoLog Journal of Logics and their Applications are those of the respective authors and contributors and not of the IfCoLog Journal of Logics and their Applications or of College Publications. Neither College Publications nor the IfCoLog Journal of Logics and their Applications make any representation, express or implied, in respect of the accuracy of the material in this journal and cannot accept any legal responsibility or liability for any errors or omissions that may be made. The reader should make his/her own evaluation as to the appropriateness or otherwise of any experimental technique described.

© Individual authors and College Publications 2017
All rights reserved.

ISBN 978-1-84890-252-7
ISSN (E) 2055-3714
ISSN (P) 2055-3706

College Publications
Scientific Director: Dov Gabbay
Managing Director: Jane Spurr

http://www.collegepublications.co.uk

Printed by Lightning Source, Milton Keynes, UK

All rights reserved. No part of this publication may be reproduced, stored in a retrieval system or transmitted in any form, or by any means, electronic, mechanical, photocopying, recording or otherwise without prior permission, in writing, from the publisher.

EDITORIAL BOARD

Editors-in-Chief
Dov M. Gabbay and Jörg Siekmann

Marcello D'Agostino
Natasha Alechina
Sandra Alves
Arnon Avron
Jan Broersen
Martin Caminada
Balder ten Cate
Agata Ciabttoni
Robin Cooper
Luis Farinas del Cerro
Esther David
Didier Dubois
PM Dung
Amy Felty
David Fernandez Duque
Jan van Eijck

Melvin Fitting
Michael Gabbay
Murdoch Gabbay
Thomas F. Gordon
Wesley H. Holliday
Sara Kalvala
Shalom Lappin
Beishui Liao
David Makinson
George Metcalfe
Claudia Nalon
Valeria de Paiva
Jeff Paris
David Pearce
Brigitte Pientka
Elaine Pimentel

Henri Prade
David Pym
Ruy de Queiroz
Ram Ramanujam
Chrtian Retoré
Ulrike Sattler
Jörg Siekmann
Jane Spurr
Kaile Su
Leon van der Torre
Yde Venema
Rineke Verbrugge
Heinrich Wansing
Jef Wijsen
John Woods
Michael Wooldridge
Anna Zamansky

Scope and Submissions

This journal considers submission in all areas of pure and applied logic, including:

- pure logical systems
- proof theory
- constructive logic
- categorical logic
- modal and temporal logic
- model theory
- recursion theory
- type theory
- nominal theory
- nonclassical logics
- nonmonotonic logic
- numerical and uncertainty reasoning
- logic and AI
- foundations of logic programming
- belief revision
- systems of knowledge and belief
- logics and semantics of programming
- specification and verification
- agent theory
- databases
- dynamic logic
- quantum logic
- algebraic logic
- logic and cognition
- probabilistic logic
- logic and networks
- neuro-logical systems
- complexity
- argumentation theory
- logic and computation
- logic and language
- logic engineering
- knowledge-based systems
- automated reasoning
- knowledge representation
- logic in hardware and VLSI
- natural language
- concurrent computation
- planning

This journal will also consider papers on the application of logic in other subject areas: philosophy, cognitive science, physics etc. provided they have some formal content.

Submissions should be sent to Jane Spurr (jane.spurr@kcl.ac.uk) as a pdf file, preferably compiled in LaTeX using the IFCoLog class file.

CONTENTS

ARTICLES

Preface to the Speical Issue on Reasoning about Preferences,
  Uncertainty, and Vagueness 1901
  Thomas Lukasiewicz, Rafael Peñaloza and Anni-Yasmin Turhan

Resolution and Clause-Learning with Restarts for Signed CNF Formulas . 1905
  David Mitchell

Fuzzy Propositional Formulas under the Stable Model Semantics 1927
  Joohyung Lee and Yi Wang

Multi-Attribute Decision Making with Weighted Description Logics 1973
  Erman Acar, Manuel Fink, Christian Meilicke, Camilo Thorne and Heiner Stuckenschmidt

Preference Inference Based on Hierarchical and
  Simple Lexiographic Models 1997
  Nic Wilson, Anne-Marie George and Barry O'Sullivan

A Semantic Approach to Combining Preference Formalisms 2039
  Alireza Ensan and Eugenia Ternovska

A Framework for Versatile Knowledge and Belief Management Operations
  in a Probabilistic Conditional Logic 2063
  Christoph Beierle, Marc Finthammer, Nico Potyka, Julian Varghese and Gabriele Kern-Isberner

Preface to the Special Issue on Reasoning about Preferences, Uncertainty, and Vagueness

Thomas Lukasiewicz
Department of Computer Science, University of Oxford, UK
`thomas.lukasiewicz@cs.ox.ac.uk`

Rafael Peñaloza
KRDB Research Centre, Free University of Bozen-Bolzano, Italy
`penaloza@inf.unibz.it`

Anni-Yasmin Turhan
Technische Universität Dresden, Germany
`turhan@inf.tu-dresden.de`

This special issue contains extended versions of a selection of the papers presented at the *First International Workshop on Reasoning about Preferences, Uncertainty, and Vagueness (PRUV 2014)*, which took place on July 23–24th, 2014, in Vienna, Austria, as part of the *Vienna Summer of Logic*.

Managing preferences, uncertainty, and vagueness has been a topic of interest for artificial intelligence (AI) since its inception. In recent years, with the availability of massive amounts of data in different repositories and the possibility of integrating and exploiting these data, technologies for managing preferences, uncertainty, and vagueness are playing a key role in other areas, most notably in databases and the (Social and/or Semantic) Web. These application areas have sparked a new wave of interest into logics capable of handling these, other kinds of meta-knowledge. Important examples are fuzzy and probabilistic approaches for description logics, rule systems for handling vagueness and uncertainty in the Semantic Web, or formalisms for handling user preferences in the context of ontological knowledge in the Social Semantic Web.

The aim of PRUV 2014 was to bring together people from different communities such as AI, the Semantic Web, or automated reasoning to name a few. The workshop started on each day with an inspiring invited talk: Tommie Meyer spoke on *Preferential Semantics as the Basis for Defeasible Reasoning* in description logics, and Gabriele Kern-Isberner investigated in her talk on *Multiple Iterated Belief Revision for Ranking Functions* richer epistemic structures like probabilities or qualitative rankings. The PRUV workshop's audience included theorists and practitioners, working on logics for reasoning about preferences, uncertainty, and vagueness. Making them aware of and fruitfully discuss the most recent application areas, new challenges, and the existing body of work on logics for reasoning about preferences, uncertainty, and vagueness was the desired outcome of the workshop. During the two days of the event, this goal was greatly surpassed.

Continuing on the success of the event, this special issue contains six papers which improve and extend those originally presented at PRUV 2014. These papers cover the different topics of interest for the workshop:

- *Resolution and Clause Learning with Restarts for Multi-Valued CNF Formulas*, by David Mitchell, proposes adaptations to the well-known methods developed in the area of (classical) propositional satisfiability to two families of signed CNF formulas. Thus, this work provides the basis for developing efficient reasoning methods for multi-valued propositional logic and other logics for vagueness.

- A different approach for handling vagueness in propositional logic is given in *Fuzzy Propositional Formulas under the Stable Model Semantics*, by Joohyung Lee and Yi Wang. In this work, the authors extend the ideas from answer set programming to handle the intermediate truth degrees from fuzzy logic. A comparison between their proposal and classical stable semantics shows that many of the desirable properties of the latter are preserved in the former.

- In *Multi-Attribute Decision Making with Weighted Description Logics*, Erman Acar, Manuel Fink, Christian Meilicke, Camilo Thorne, and Heiner Stuckenschmidt show how to compactly represent attribute preferences and choices using a description logic knowledge base extended with utility values. Complementing the theoretical results, the work presents an empirical evaluation of an implemented tool and its application to a use case based on DBpedia.

- The following paper, *Preference Inference Based on Hierarchical and Simple Lexicographic Models*, by Nic Wilson, Anne-Marie George, and Barry O'Sullivan, deals with the problem of deducing new preference relationships from a set of known preferences, using a lexicographic model. The authors show that deciding whether a new preference statement can be decided from previous observations is infeasible in general. However, they provide different restricted settings that retain feasibility.

- As a means to combine preference descriptions, possibly specified in different languages, *A Semantic Approach to Combining Preference Formalisms*, by Alireza Ensan and Eugenia Ternovska, proposes a modular framework with combination operations. They show the effectiveness of their formalism by combining different languages explicitly.

- Finally, *A Framework for Versatile Knowledge and Belief Management Operations in a Probabilistic Conditional Logic*, by Christoph Beierle, Marc Finthammer, Nico Potyka, Julian Varghese, and Gabriele Kern-Isberner, deals with notions of uncertainty. In this work, the authors propose using probabilistic conditional logic with the principle of maximum entropy for modelling intelligent agents. Interestingly, an implementation of this approach is used to model data and expert knowledge from oncology.

We would like to thank all authors for their contributions to this special issue and all referees (Mario Alviano, Daniel Borchmann, Marco Cerami, Simona Colucci, Fabio G. Cozman, Dragan Doder, Souhila Kaci, Enrico Malizia, Felip Manya, Maria Vanina Martinez, Livia Predoiu, Fabrizio Riguzzi, Ganesh Ram Santhanam, Francesco Santini, Steven Schockaert, Gerardo I. Simari, Matthias Thimm, Oana Tifrea-Marciuska, and Kewen Wang) for their timely expertise in carefully reviewing the contributions.

Thomas Lukasiewicz
Rafael Peñaloza
Anni-Yasmin Turhan

Resolution and Clause-Learning with Restarts for Signed CNF Formulas

David Mitchell
Simon Fraser University
mitchell@cs.sfu.ca

Abstract

Motivated by the question of how to efficiently do model finding or theorem proving for multi-valued logics, we study the relative reasoning power of resolution proofs and a natural family of model-finding algorithms for Signed CNF Formulas. The conflict-driven clause learning (CDCL) algorithm for SAT is the basis of model finding software systems (SAT solvers) that have impressive performance on many families of propositional formulas. CDCL with restarts (CDCL-R) has been shown to have essentially the same reasoning power as unrestricted propositional resolution. More precisely, they p-simulate each other. We show that this property generalizes to two families of Signed CNF formulas, those with unrestricted signs, and those where the truth value set is a lattice and all signs are regular. We show that a natural generalization of CDCL-R to these formulas has essentially the same reasoning power as natural generalizations of resolution found in the literature. Moreover, the algorithm efficiently simulates bounded width resolution in these systems. These families of signed formulas are possible reduction targets for a number of multi-valued logics, and thus this algorithm has potential as a basis for efficient implemented reasoning systems for many multi-valued logics.

1 Introduction

Multi-valued logics are among the most established and widely studied formalisms for reasoning with uncertainty. In this paper we consider Signed CNF formulas, as defined, for example, in [10, 6], and study the relative reasoning power of resolution proofs and a natural class of algorithms to decide their satisfiability.

The dominant algorithm in modern implemented model-finders for propositional satisfiability (SAT solvers) is the conflict-driven clause learning algorithm (CDCL)

This research was supported in part by a Natural Sciences and Engineering Research Council of Canada (NSERC) Discovery Grant.

introduced in [15], with restarts [13], here denoted CDCL-R. Good solvers based on CDCL-R have remarkable performance on many families of formulas. Consequently, many practical reasoning tasks are carried out by reduction to propositional CNF, or by adaptation of CDCL to other families of formulas. This leads us consider whether we could best obtain effective model finders or theorem provers for multi-valued logics by reduction to SAT or by adapting the CDCL algorithm to a multi-valued context. The goal of this paper is to contribute to our understanding of the potential of adaptations of CDCL-R to the multi-valued case.

Validity or satisfiability of many multi-valued and fuzzy logics (as well as annotated logics, and others) can be reduced naturally to satisfiability of signed CNF formulas [9]. Natural versions of resolution for signed CNF formulas have been presented in the literature, and it is possible to produce natural variants of the CDCL algorithm for these formulas as well. We study two particular families of signed formulas. One family, denoted here MV-CNF, is a very slight restriction on signed CNF formulas with unrestricted signs; the other family consists of Regular CNF formulas with a domain that is a lattice, here denoted Reg-CNF. Corresponding versions of binary resolution give us proof systems for each family, here denoted MV-RES and Reg-RES, respectively. We give an abstract signed version of the CDCL-R algorithm, denoted MV-CDCL-R, which can be instantiated for either family of formulas and many others.

Here, as is generally the case in proof complexity, we measure the reasoning power of a system in terms of minimum proof length. For an algorithm such as CDCL-R, the corresponding notion is minimum length of execution given optimum decision and restart policies. It has been shown that CDCL-R with unlimited restarts has, up to a small polynomial factor, the same reasoning power as unrestricted propositional resolution [18]. (It is not currently known whether restarts are essential or not. See the brief discussion in Section 7.) In [2] it was shown that that CDCL-R can efficiently refute CNF formulas that have bounded-width refutations.

The main purpose of this paper is to show that these properties generalize to the multi-valued systems in question. The main result is that MV-RES and the MV-CDCL-R algorithm for MV-CNF formulas are of essentially the same efficiency: For unsatisfiable MV-CDCL-R formulas, the minimum size proofs of unsatisfiability in the two systems differ in size by at most a small polynomial. Formally, the two systems are said to p-simulate each other, meaning there are polynomial-time functions mapping any proof in one system to a proof in the other. The same property is shown for Reg-RES and the corresponding Reg-CNF version of MV-CDCL-R.

The largest part of the proof consists of showing that the MV-CDCL-R algorithm can efficiently simulate arbitrary resolution refutations. This simulation yields the following concrete bounds. A resolution refutation is of size r and width w if it has r clauses and the largest clause has at most w literals. We show that, if Γ is an unsatisfiable MV-CNF formula or Reg-CNF formula with s literals, over n atoms, and has a resolution refutation (in the corresponding resolution proof system) of size r and width w, then there is an execution of MV-CDCL-R that refutes Γ, with the following properties:

1. It implicitly generates a resolution refutation of size $O(wn^2r)$.

2. It can be executed in time $O(wn^2sr)$;

It follows that, for any fixed w, MV-CDCL-R can refute formulas that have width-w resolution refutations in time $O(n^{w+2}s)$, with an implicit resolution refutation constructed of size $O(n^{w+2})$.

Our proof is an adaptation of that in [18], with parts influenced by [2], although our presentation is distinct. Rather than proceed from a detailed examination of CDCL-R, we proceed from the key properties of resolution proofs, to a simplified algorithm that performs the main reasoning in CDCL-R, and then to a highly abstracted version of CDCL-R. This emphasizes the properties in the proofs and in the algorithm that are most relevant, in particular those involving "empowering clauses". Intuitively, a clause C is empowering for a set Γ of clauses if more can be proven by unit resolution from $\Gamma \cup \{C\}$ than from Γ alone. The key properties that are exploited by the proofs are that:

1. Non-trivial resolution refutations can be decomposed into a sequence of derivations of empowering clauses;

2. The clauses derived by the clause learning mechanism in the CDCL algorithm are empowering clauses.

Given a resolution refutation Π of clause set Γ, it is possible to use decision and restart policies to direct CDCL-R to efficiently construct a refutation of Γ that is not much longer than Π. This is done by repeatedly finding an empowering clause from Π, and then causing CDCL-R to generate one or more learned clauses that are "as good as" (in a sense to be made precise) that clause.

The main technical result in [18] is that, for any formula with a resolution refutation Π of length r, CDCL-R can implicitly generate a resolution refutation of length $O(n^4r)$. By slightly more careful counting, we obtain size $O(wn^2r)$, where w is the width of the given refutation Π. For general refutations, with no restriction on clause width, this gives us size $O(n^3r)$.

In [2] it is shown that CDCL-R, with sufficiently many random decisions and sufficiently frequent restarts, with high probability refutes any formula having a width-w resolution refutation with using at most $O(n^{2w+2})$ conflicts, and therefore an implicit resolution refutation of size $O(n^{2w+3})$. For the restriction to formulas with refutations of width bounded by some fixed w, our deterministic bound gives a resolution refutation size of $O(n^2 r)$ which is $O(n^{w+2})$ because $r = O(n^w)$.

The organization of the paper is as follows. In Section 2 we define signed CNF, MV-CNF and Reg-CNF formulas, along with binary resolution rules for these formulas. The properties of resolution proofs that are central to the proof are defined and established in Section 3. In Section 4 we describe an algorithm that embodies the core reasoning in MV-CDCL-R (as well as standard CDCL-R), while Section 5 shows that repeated calls to this algorithm can refute formulas almost as efficiently as unrestricted resolution. In Section 6 we give our MV-CDCL-R algorithm, define p-simulation and give the main theorem. We briefly discuss applicability issues in Section 7, and conclude in Section 8.

A preliminary version of this paper appeared as [17]. The present version corrects notational problems, simplifies and clarifies the presentation, gives tighter simulations and adds the discussion on applicability.

2 Signed CNF Formulas and Resolution

Let T be a finite set of truth values, $\mathcal{P}$ a countably infinite set of multi-valued atoms and $\prec$ a partial order on T. We assume throughout that T is fixed, so the size of T is a constant in complexity analyses. Signed CNF formulas for T are constructed from literals of the form $p \mathop{\in} S$, where S is a non-empty proper subset of T and $p \in \mathcal{P}$. We will use S and R for sets of truth values in literals, and l, often with subscripts, for literals. Complementation of sets of truth values is taken with respect to T, so $\overline{S}$ denotes $\{a \in T \mid a \notin S\}$. If l is the literal $p \mathop{\in} S$, its complement $\overline{l}$ is the literal $p \mathop{\in} \overline{S}$. A clause is a disjunction of literals, and a formula is a conjunction of clauses. When convenient we identify clauses with sets of literals, and formulas with sets of clauses.

Remark 1. *In the literature on signed formulas, literals are typically written $S{:}p$, but we prefer the readability of a set-like notation such as $p \mathop{\in} S$. Formally S is a sequence of constant symbols denoting elements of T and enumerating the set of values p may take. Following usual practice in the literature, we overload S and use it both for the set of truth values and the string representing this set.*

An assignment τ for formula Γ and truth value set T is a function mapping the propositional atoms of Γ to T. Our assignments will often be partial. Assignment τ satisfies a literal $p \mathop{\in} S$ if $\tau(p) \in S$, satisfies a clause C if it satisfies at least one literal in C, and satisfies CNF formula Γ if it satisfies every clause in Γ. If X is a literal, clause, or formula, we write $\tau \models X$ to indicate that τ satisfies X. A formula or set of formulas Φ entails a formula Γ, written $\Phi \models \Gamma$, if every assignment that satisfies each formula in Φ also satisfies Γ. In particular, if $S \subset R$, then every assignment that satisfies $p{\in}S$ also satisfies $p{\in}R$. We say $p{\in}S$ entails $p{\in}R$, and write $p{\in}S \models p{\in}R$.

The connectives in signed formulas are classical two-valued connectives. Signed CNF formulas are intended as a reduction target so, for example, to find a model of a multi-valued formula ϕ, we would transform it into an appropriate signed CNF formula Γ_ϕ, and then run a model finder for signed CNF. It is the responsibility of the transformation to correctly handle the semantics of non-classical connectives in the source logic. A general, linear-time transformation of formulas from an arbitrary finitely valued logic to Signed CNF is described in [9]

A number of variants of this basic logic have been studied. Typical variants restrict the allowed literals or impose structure on the truth value set T. We will explicitly examine two, although the method can be adapted to some others. (Unfortunately, terminology is not uniform in the literature, so our terms may correspond only roughly to those in some other papers.)

1. **Many-Valued CNF (MV-CNF):** Signed CNF formulas as just described, with the restriction that each atom p occurs in at most one literal in any clause. The order relation $\prec$ plays no role.

2. **Regular CNF over a lattice (Reg-CNF):** Let $\langle T, \prec \rangle$ be a lattice. We call a literal $p{\in}S$ regular for $\langle T, \prec \rangle$ iff S is either the upset $\uparrow a = \{b \in T \mid a \preceq b\}$, or the downset $\downarrow a = \{b \in T \mid b \preceq a\}$ of some $a \in T$. A formula Γ is regular if every literal in Γ is regular.

When making statements that apply to both families of formulas, we sometimes use the terms "signed CNF" or "signed formula", rather than explicitly saying "MV-CNF or Reg-CNF formula".

Example 1. *Consider the signed formulas with $T = \{0, 1\}$. The MV-CNF version is equivalent to the classical case, as is the Reg-CNF version with $0 \prec 1$.*

Example 2. *MV-CNF formulas for $\langle T, \prec \rangle$, with $T = \{0, \frac{1}{d-1}, \frac{2}{d-1} \ldots 1\}$, for some $d \in \mathbb{N}$, and $\prec$ the standard order on $\mathbb{Q}$, are a natural reduction target for commonly used multi-valued logics, including the finite-valued Łukasiewicz logics. In the analogous Reg-CNF version, literals are restricted to those equivalent to $p < a$ and $p > a$, for some $a \in S$. In MV-RES reasoning, the order on T is ignored.*

2.1 Signed Resolution

Let signed binary resolution be the following derivation rule (where A and B are arbitrary disjunctions).

$$\frac{(p{\in}S \ \vee \ A) \quad (p{\in}R \ \vee \ B)}{(p{\in}(S\cap R) \ \vee \ A \ \vee \ B)} \tag{1}$$

We say the two antecedent (top) clauses in (1) were resolved on p to produce the resolvent clause. Two literals $p{\in}S$ and $p{\in}R$ clash if $S \neq R$. If $R \cap S$ is empty, we say the clash is annihilating, and otherwise we call the literal $p{\in}(R \cap S)$ the residue. A pair of clashing literals that are annihilating are inconsistent.

As in the classical case, a resolution derivation Π of clause C from a set Γ of clauses is a sequence of clauses $\langle C_1, \ldots, C_r \rangle$, where each C_i is either in Γ or derived from two earlier clauses in Π by the resolution rule, and $C_r = C$. The length, or size, of the derivation is the number of clauses r. A resolution refutation of Γ is a resolution derivation from Γ of the empty clause, denoted $\square$.

A resolution rule is sound and refutation complete for a family of formulas if, for every formula Γ in the family, Γ is unsatisfiable iff it has a refutation constructed using the rule. Rule (1) is the basis of resolution proof systems for our two families of formulas, but we need different variants to obtain a sound and complete proof system for each family.

Resolution for MV-CNF: We obtain a sound and refutation-complete proof system by providing rule (1) with implicit merging and annihilation [10]. That is:

1. (Merging) If two literals $p{\in}S$ and $p{\in}R$ with the same atom p occur in the resolvent, they are replaced by $p{\in}(S \cup R)$.
2. (Annihilation) Whenever $(S \cap R)$ is empty, the "false literal" $p{\in}\varnothing$ is omitted from the resolvent.

We denote the resulting system MV-RES.

Resolution for Reg-CNF: The following restricted signed resolution rule is sound and complete for regular formulas over a lattice [5].

$$\frac{(p{\in}{\uparrow}a \ \vee \ A) \quad (p{\in}{\downarrow}b \ \vee \ B)}{(A \vee B)} \quad \text{provided} \ \ a \not\leq b. \tag{2}$$

We denote the resulting system by Reg-RES.

The differences in the properties of the resolution rules of these two systems are the main thing to be delt with in adapting the CDCL-R algorithm and the proofs from [18] to our setting. In MV-RES, literals have complements, and no atom appears in multiple literals of any clause, but resolution is not annihilating and we must allow for the residues. In Reg-RES, resolution is annihilating, but literals need not have complements (more precisely, the complement of a regular literal may not be a regular). Literal complementation and the annihilating property of classical resolution both play central roles in the classical CDCL-R algorithm.

3 Empowering and Absorbed Clauses

The efficient simulation of resolution by CDCL-R relies on a property of resolution refutations involving unit resolution. A unit clause is a clause with exactly one literal, and unit resolution is application of the resolution rule when at least one antecedent is a unit clause. We write $\Gamma \vdash^u (l)$, or simply $\Gamma \vdash^u l$, if (l) can be derived from Γ by unit resolution alone, and write $\Gamma \vdash^u \square$ if there is a refutation of Γ using only unit resolution. As in the classical case, with appropriate data structures it is possible to check if $\Gamma \vdash^u l$ or $\Gamma \vdash^u \square$ in linear time. (A proof of this is provided in Appendix A).

We will use certain sets of literals that are inconsistent with a given clause. For a set or sequence L of literals, we denote by $\widetilde{L}$ the set of literals which are the complements of literals in L. In particular, if C is a clause, then $\widetilde{C}$ is the set of complements of literals in C. Semantically, we view $\widetilde{C}$ as a conjunction of literals that is equivalent to $\neg C$. For Reg-CNF we must ensure all literals are regular, but even if C contains only regular literals $\widetilde{C}$ may not. We define a set $\widehat{L}$ as follows. If $L = l_1, l_2, \ldots l_k$ is a sequence of literals, $\widehat{L}$ denotes the set of all size-k sequences of the form $L' = l'_1, \ldots l'_k$, where each l'_i is a regular literal that is inconsistent with l_i. If L is empty then $\widehat{L}$ contains only the empty sequence.

If Γ is a set of clauses and L a set or sequence of literals, we may write Γ, L as an abbreviation for $\Gamma \cup \{(l) \mid l \in L\}$. If C is a clause of size k then $\Gamma, \widetilde{C}$ is the set of all clauses from Γ plus k unit clauses corresponding to the k literals in C. Thus $\Gamma, \widetilde{C} \vdash^u \square$ indicates, intuitively, that the restriction of Γ obtained by setting all literals of C false can be refuted by unit resolution.

Definition 1 (Empowering and Absorbed Clauses). *Let Γ be a set of clauses and C a clause with $\Gamma \models C$. For MV-CNF formula Γ we say C is l-empowering for Γ iff $C = (A \vee l)$ and*

1. $\Gamma, \widetilde{C} \vdash^u \square$;
2. $\Gamma, \widetilde{A} \not\vdash^u \square$;

3. For any literal l', if $l' \models l$, then $\Gamma, \widetilde{A} \not\vdash^\mu l'$.

For Γ a Reg-CNF formula, we say that C is l-empowering for Γ iff $C = (A \lor l)$ and

1a. $\Gamma, C' \vdash^u \square$, for some $C' \in \widehat{C}$;

2a. $\Gamma, A' \not\vdash^\mu \square$ for each $A' \in \widehat{A}$;

3a. $\Gamma, A' \not\vdash^\mu l'$ for each $A' \in \widehat{A}$, and each l' such that $l' \not\models l$;

Clause C is empowering for Γ if it is l-empowering for some $l \in C$, and is absorbed by Γ otherwise.

Lemma 1 (Existence of Empowering Clauses). *Let Γ be a set of signed clauses for which $\Gamma \not\vdash^u \square$. If Π is a signed resolution refutation of Γ, then Π contains a clause that is empowering for Γ.*

Proof. Let C be the first clause in Π that does not satisfy condition 1 (or 1a, respectively) of Definition 1. Such a clause exists, because $\square$ suffices if no earlier clause does. C is the resolvent of two earlier clauses of Π, say $C_1 = (p \in S_1 \lor A_1)$, and $C_2 = (p \in S_2 \lor A_2)$, where $p \in S_1$ and $p \in S_2$ clash. One of C_1 or C_2 is empowering for Γ. To see this, first observe that both are logically implied by Γ, because they are in Π and signed resolution is sound, and both satisfy condition 1 (respectively, 1a) of Definition 1 by choice of C.

We complete the argument for MV-RES as follows. Both C_1 and C_2 satisfy condition 2 of Definition 1, because $C = (p \in (S_1 \cap S_2) \lor A_1 \lor A_2)$, so if $\Gamma, \widetilde{A_1} \vdash^u \square$ or $\Gamma, \widetilde{A_2} \vdash^u \square$ then $\Gamma, \widetilde{C} \vdash^u \square$, contradicting choice of C. Now, suppose both C_1 and C_2 fail condition 3 of Definition 1. That is, for some $R_1 \subset S_1$ and $R_2 \subset S_2$, we have $\Gamma, \widetilde{A_1} \vdash^u (p \in R_1)$ and $\Gamma, \widetilde{A_2} \vdash^u (p \in R_2)$. Resolving $(p \in R_2)$ and $(p \in R_1)$ produces the unit clause $(p \in (R_1 \cap R_2))$, where $(R_1 \cap R_2) \subset (S_1 \cap S_2)$. If follows that $\Gamma, \widetilde{C} \vdash^u \square$, because $p \in \overline{(S_1 \cap S_2)}$ is in $\widehat{C}$ and we can resolve it with $p \in (R_1 \cap R_2)$ to obtain $\square$. This again contradicts choice of C, so at least one of C_1 or C_2 is empowering for Γ.

The completion for Reg-RES is just a variation. $C = (A_1 \lor A_2)$, because Reg-RES has no residuals, so if $\Gamma, A_1' \vdash^u \square$ for some $A_1' \in \widehat{A_1}$ then $\Gamma, C' \vdash^u \square$ for some $C' \in \widehat{C}$ (because $A_1 \subset C$) contradicting choice of C. By the symmetric argument, there is no $A_2' \in \widehat{A_2}$ for which $\Gamma, A_2' \vdash^u \square$. So, both C_1 and C_2 satisfy condition 2a. Now, suppose both C_1 and C_2 fail condition 3. Then, for some $A_1' \in \widehat{A_1}$, $A_2' \in \widehat{A_2}$, $R_1 \subset S_1$ and $R_2 \subset S_2$, we have that $\Gamma, A_1' \vdash^u (p \in R_1)$ and $\Gamma, A_2' \vdash^u (p \in R_2)$. Because Reg-RES has no residuals, we know $S_1 \cap S_2 = \varnothing$, so we also have that $R_1 \cap R_2 = \varnothing$ and $(p \in R_1)$ and $(p \in R_2)$ can be resolved to produce $\square$. So if $C' \in \widehat{C}$ then $\Gamma, C' \vdash^u \square$, again contradicting the choice of C. So either C_1 or C_2 is empowering for Γ. $\square$

4 Probing with Learning

The core of the CDCL algorithm can be viewed as a back-and-forth between two tightly related processes, one which guesses at partial assignments, and one which derives new clauses based on these guesses and what follows from them by unit propagation. We first consider an algorithm, which we call Probe-and-Learn, that embodies one stage of this interaction. Describing the algorithm requires some terminology.

For signed formulas, the guesses (called "decisions" in the SAT literature) involve restrictions on assignments, rather than absolute assignments. We will make use of certain sequences of restrictions.

Definition 2 (Proper Restriction Sequence). *A proper restriction sequence (or just "restriction sequence") δ for T and Γ is a sequence $\delta = \langle l_1, l_2, \ldots l_k \rangle$ of distinct literals for T such that:*

1. *If δ contains a literal $p {\in} S$ then Γ has a literal with the atom p;*

2. *The set of literals in δ is satisfiable;*

3. *If l_i is $p {\in} S$, then $\bigcap \{R \mid j < i \text{ and } l_j = p {\in} R\} \cap \overline{S} \neq \varnothing$*

For any non-empty restriction sequence $\delta = \langle l_1, \ldots l_{k-1}, l_k \rangle$, we denote by δ^- the maximal proper prefix $\delta^- = \langle l_1, \ldots l_{k-1} \rangle$.

We will use δ and α for decision sequences. Condition 3 of the definition ensures that each successive literal in the restriction sequence further restricts the allowed assignments.

We say that assignment τ is consistent with restriction sequence δ if no literal in δ is inconsistent with a literal in τ. For literal l and restriction sequence δ, we may say that "δ makes l false" if l is not satisfied by any assignment consistent with δ, and that "δ makes l true" if every assignment consistent with δ satisfies l.

Unit propagation is a key component of CDCL algorithms, and in particular of the "back-and-forth" process embodied in Probe-and-Learn. We will define unit propagation for signed formulas, as used in our algorithm, in terms of restriction sequences.

Definition 3 ($\mathrm{UP}(\Gamma, \delta)$). *For any clause set Γ and restriction sequence δ for Γ, we denote by $\mathrm{UP}(\Gamma, \delta)$ the restriction sequence δ' defined by the fixpoint of the following operation:*

> *If Γ contains a clause $C = (l \vee B)$ where δ makes every literal of B false, but does not make l either true or false, extend δ with l.*

Algorithm 1: Probe-and-Learn

Input: Clause set Γ; truth value set $\langle T, \prec \rangle$; restriction sequence δ.
Output: Clause set Γ'; restriction sequence δ'.

1   $\alpha \leftarrow$ a minimal extension of δ such that either $\mathrm{UP}(\Gamma, \alpha) \models \Gamma$ or $\Gamma, \alpha \vdash^u \square$
2   **while** $\Gamma, \alpha \vdash^u \square$ and $\alpha \neq \langle \rangle$ **do**
3      $\alpha, C \leftarrow$ Handle-Conflict(Γ, α)
4      $\Gamma \leftarrow \Gamma \cup \{C\}$
5   **end**
6   **return** Γ, α

Unit propagation corresponds to unit resolution, in a context where we are interested in collecting implied restrictions on truth assignments rather than derived unit clauses. In particular, the restriction sequence $\mathrm{UP}(\Gamma, \delta)$ makes a clause of Γ false if and only if $\Gamma, \delta \vdash^u \square$.

A second process, closely related to unit propagation, involves derivation of clauses called "asserting clauses". CDCL-R proofs of unsatisfiability are constructed by a sequence of asserting clause derivations.

Definition 4 (Asserting Clause; Conflict Clause). *Clause C is an asserting clause for signed CNF formula Γ and restriction sequence δ iff*

1. $\Gamma, \widetilde{C} \vdash^u \square$ *if Γ is MV-CNF; $\Gamma, A \vdash^u \square$ for each $A \in \widehat{C}$ if Γ is Reg-CNF;*

2. *For each literal $l \in C$, there is a literal l' with $l' \models \bar{l}$ and $\Gamma, \delta \vdash^u l'$;*

3. *For exactly one literal $l \in C$, there is no literal l' with $l' \models \bar{l}$ and $\Gamma, \delta^- \vdash^u l'$.*

C is a conflict clause for Γ and δ if it satisfies conditions 1 and 2.

The Probe-and-Learn algorithm is presented in Algorithm 1. It is parameterized by T and $\prec$ so we don't need to present distinct versions for MV-CNF and Reg-CNF. The differences only affect low level details involving operations on literals. In analyses, we take the parameter $\langle T, \prec \rangle$ to be fixed, and allow the other two arguments, the clause set Γ and restriction sequence δ, to vary.

Probe-and-Learn extends δ to a minimal extension α of δ for which unit propagation either produces a satisfying assignment or makes a clause false. In the former case, Γ and the satisfying assignment α are returned. In the latter case Handle-Conflict is executed. Handle-conflict returns a proper prefix of α (which becomes the new value of α) and a clause C which is added to Γ ("learned").

We require Handle-Conflict to satisfy the following correctness property. If the call Handle-Conflict(Γ, α) returns α', C then

1. α' is a proper prefix of α;

2. If α' is empty, then C is $\square$;

3. If α' is not empty, then C is an asserting clause for Γ and UP(Γ,α').

If C is an asserting clause for Γ and α', then UP($\Gamma \cup \{C\}),\alpha'$) will be a proper extension of UP(Γ,α), and it is possible for this propagation to reach a new conflict, after which a new asserting clause may be derived. The loop on lines 2-5 of Probe-and-Learn repeats this process until unit propagation no longer produces a conflict (or Γ is proven unsatisfiable). Probe-and-Learn returns the resulting clause set and restriction sequence.

For correctness, there is no restriction on the method by which Handle-Conflict generates C and α. For the p-simulation results of Section 6.1, Handle-Conflict must run in polynomial time. For the concrete simulation bounds of Section 5 and Section 6, it must run in time $O(ns)$, where n is the number of atoms and s is the total number of literal occurrences in the clause set, and there must be a corresponding (possibly implicit) resolution derivation of each asserting clause.

For simplicity of the remainder of the presentation, we will assume that Handle-Conflict is implemented by an algorithm analogous to the standard method that forms the basis of conflict clause derivation in almost all CDCL SAT solvers. We now describe this algorithm, and show that it does run in the require time bound and construct an appropriate resolution derivation.

4.1 Asserting Clause Derivation

An execution of Handle-Conflict(Γ,α) must, except in the case it finds a satisfying assignment, return an asserting clause for a proper prefix of α. The standard methods for this in CDCL SAT solvers involve a resolution derivation closely connected to the unit propagation sequence that establishes a conflict. (The method may either be implemented based on resolution, as we describe below, or the "implication graph" [15].) A generalized version of this process can be used in our many-valued Handle-Conflict procedure. We describe a particular version, the so-called "1UIP asserting clause" derivation. Most CDCL SAT solvers use a refined version of this.

We make the assumption, consistent with standard practice, that UP(Γ,δ) is computed incrementally according to the order of literals in δ. That is, if $\delta = l_1, l_2, l_3...$, we first extend δ by computing UP(Γ,l_1), then extend UP(Γ,l_1) by any additional literals in UP($\Gamma,\langle l_1, l_2\rangle$), etc., so that the last literals in UP(Γ,δ) are those that are not also in UP(Γ,δ^-). The elements of δ in UP(Γ,δ) are called "decisions" (they are the guesses), and the others are there because they are implied.

We obtain the desired asserting clause by means of a resolution derivation constructed as follows. Let $L = l_1, l_2, \ldots, l_d, \ldots, l_k$ be the literals of UP(Γ,δ) in order, and l_d be the last literal of L that is in δ (i.e., the last decision literal in L). We associate to each literal l_i in $l_{d+1}, \ldots, l_r$ a pair of clauses B_i and a C_i. For each i in $d+1, \ldots, r$, let B_i be a clause of Γ that, when restricted by $l_1, \ldots l_{i-1}$, is the unit clause (l_i). Such a B_i must exist, since l_i was obtained by unit propagation.

We define the C_i by induction in reverse order as follows. Let C_{r+1} be a clause of Γ that is made false by L. For each i in $r \ldots d+1$ (proceeding in that order) let C_i be the resolvent of B_i and C_{i+1} if they are resolvable, and C_{i+1} if they are not. It is clear that each clause C_i in the sequence can be derived from Γ by resolution, and that the number of derived clauses in this derivation is at most the length of the sequence L. Let j be the largest index in $d+1, \ldots r$ for which C_j contains only one literal that is inconsistent with a literal in the sequence $l_j, \ldots l_r$. (Certainly $j = d+1$ will work, if no larger value does.) The clause C_j, which is known in the SAT literature as the 1UIP clause (for "first unique implication point"), is the clause to be returned by Handle-Conflict. The restriction sequence to be returned by Handle-Conflict is the least prefix δ' of δ such that UP(Γ,δ') makes C_j a unit clause.

Lemma 2. *Each derived clause in the derivation of the 1UIP clause is a conflict clause for δ and Γ, and the clause returned by HandleConflict is an asserting clause for δ and Γ.*

Proof. We give a proof for MV-RES. The proof for Reg-RES is almost identical. C_{r+1} satisfies property 1 of Definition 4, because $C_{r+1} \in \Gamma$ and trivially for any C we have $\{C\}, \widetilde{C} \vDash^u \square$, so $\Gamma, \widetilde{C_{r+1}} \vDash^u \square$. Also, by choice C_{r+1} is made false by UP(Γ,δ), so it satisfies property 2. So C_{r+1} is a conflict clause. Now, assume that some C_{i+1} is a conflict clause. $\delta, \widetilde{C_i} \vDash^u \square$ because C_i contains every literal of $B_i \cup C_{i+1}$ except for possibly l_i and some literal l that clashes with it. So either $\widetilde{C_i}$ makes C_{i+1} false or it makes C_{i+1} and B_i clashing unit clauses. UP(Γ,δ) makes C_i false because each of its literals is either in C_{i+1} or in $B_i \setminus l_i$, both of which are made false by UP(Γ,δ). Thus, all the C_i are conflict clauses. C_j satisfies property 3 of Definition 4 by choice, so is an asserting clause. □

4.2 Complexity of Probe-and-Learn

In all complexity analyses, given a formula Γ, n will be the number of distinct atoms, the size s the number of literal occurrences, and $|\Gamma|$ the number of clauses.

Lemma 3. *Let Γ be a formula of size s over n atoms. If δ is a restriction sequence for Γ s.t. $\Gamma, \delta \vDash^u \square$, and Probe-and-Learn($\Gamma, \delta$) returns Γ', δ', then*

1. The number of conflicts generated during execution is $|\Gamma'| - |\Gamma| < |\delta|$;

2. The number of clauses derived in conflict clause derivation is at most $|\delta|O(n)$;

3. The execution can be carried out in time $|\delta|O(ns)$.

Proof. Each iteration of the body of the loop performs unit propagation and executes Handle-Conflict, which performs the asserting clause derivation. On each iteration of the loop, except possibly the terminating iteration in the case that a satisfying assignment is found, δ is set to a proper prefix of its previous value. Therefore, the number of iterations, the number of conflicts, and the number of asserting clauses added to Γ, are all at most $|\delta|$. The number of derivation steps during one asserting clause derivation is at most the maximum number of literals in a restriction sequence, which is $|T|n = O(n)$. Each resolution derivation step can be carried out in time $O(n)$. The time spent by Handle-Conflict is the time to do one unit propagation and up to n resolution steps, so is $O(s + n^2) = O(ns)$. So the total time for Probe-and-Learn is $|\delta|O(ns) = O(n^2 s)$. □

5 Simulating Resolution with Probe-and-Learn

We now show that, if Π is a resolution refutation of MV-CNF or Reg-CNF formula Γ, there is a sequence of calls to Probe-and-Learn that refutes Γ in time polynomial in the combined size of Π and Γ. We begin by showing that any empowering clause can be absorbed by a sequence of calls to Probe-and-Learn. Througout, n is the number of distinct atoms and s the number of literal occurrences, of formula Γ.

Lemma 4. *Suppose $C = (A \vee l)$ is l-empowering for clause set Γ. Then there is a sequence of calls to Probe-and-Learn that generates a superset Γ' of Γ such that C is not l-empowering for Γ', and the following properties hold.*

1. The number of calls to Probe-and-Learn is $O(n)$;

2. The size of the underlying resolution derivation is $O(n^2)$

3. The entire computation can be carried out in time $O(n^2 s)$;

Proof. We state the proof for Reg-RES; that for MV-RES is similar. Let δ be a restriction sequence consisting of the literals of some element of $\widehat{A}$, in any order, followed by a literal that is inconsistent with l. We construct a sequence $\Gamma_0, \Gamma_1, \ldots \Gamma_r$ of increasing (by set inclusion) clause sets with $\Gamma = \Gamma_0$, such that C is not l-empowering for Γ_r, and set $\Gamma' = \Gamma_r$. We obtain Γ_{i+1} by setting $\Gamma_{i+1}, \alpha = $ Probe-and-Learn(Γ_i, δ), and ignoring α. Each execution of Probe-and-Learn involves one or more executions of Handle-Conflict(Γ_i, δ'), where δ' is a prefix (not necessarily proper) of δ. We consider the sequence of all calls to Handle-Conflict, over however many calls to Probe-and-Learn are made. Each execution of Handle-Conflict returns a pair $\langle \alpha, B \rangle$, where α is a proper prefix of δ and B is an asserting clause for UP(Γ_i, α) and Γ_i. The "learned clause" B is added to Γ_i. Clause B, when restricted by UP(Γ_i, δ^-) is a unit clause. By construction, for each call to Handle-Conflict this implied unit clause will contain a distinct literal on the same atom as l. Therefore, after $O(n)$ calls to Handle-Conflict, Γ_i must contain either $\square$ a clause B for which the implied unit clause entails l. In either case, C is not l-empowering for Γ_i. Each call to Handle-Conflict requires time at most $O(ns)$, and derives $O(n)$ clauses. $\square$

Lemma 5. *If C is empowering for Γ, and C is of width (size) w, then there is a sequence of calls to Probe-and-Learn that generates a superset Γ' of Γ which absorbs C, and such that the following hold.*

1. *The number of calls to Probe-and-Learn is $O(wn)$;*
2. *The size of the underlying resolution derivation is $O(wn^2)$;*
3. *The entire computation can be executed in time $O(wn^2 s)$.*

Proof. Apply Lemma 4 for each literal $l \in C$. $\square$

To see that an appropriate sequence of calls to Probe-and-Learn can refute Γ in a number of steps not much greater than the size of any given refutation, we identify a sequence of empowering clauses, and absorb each. By "refutes Γ", we mean that it produces a set Γ' of clauses, each of which is implied by Γ, and containing $\square$.

Lemma 6. *Let Γ be a set of signed clauses and Π a signed resolution refutation of Γ of size at most r clauses, in which no clause has width greater than w. Then there is a sequence of calls to Probe-and-Learn that refutes Γ, such that the following hold.*

1. *The number of calls to Probe-and-Learn in the is $O(wnr)$;*
2. *The size of the underlying resolution refutation is $O(wn^2 r)$;*
3. *The entire computation can be executed in time $O(wn^2 sr)$.*

Proof. We generate a sequence $\Gamma_0 \ldots \Gamma_k$ of supersets of Γ, where $\Gamma_0 = \Gamma$, $k \leq r$, and $\square \in \Gamma_k$, as follows. If $\Gamma_i \vdash^u \square$ we are done. Otherwise, let C be the first clause in Π that is empowering for Γ_i. Lemma 1 ensures such a clause exists. By Lemma 5 there is a sequence of calls to Probe-and-Learn that generates a superset of Γ_i for which C is absorbed. Let this set be Γ_{i+1}. The claims follow from Lemma 5 and the fact that the number of clauses to be absorbed is at most r. $\square$

Algorithm 2: Multi-Valued CDCL with Restarts (MV-CDCL-R)

Input: Finite set Φ of signed clauses; truth value set $\langle T, \prec \rangle$
1. **Output:** SAT or UNSAT
2. $\Gamma \leftarrow \Phi$ // Clause set, initialized to the input clauses
3. $\delta \leftarrow \langle \rangle$ // Restriction sequence, initialized to empty
4. **repeat**
5. $\quad \Gamma, \delta \leftarrow$ **Probe-and-Learn**$(\Gamma, \delta, \langle T, \prec \rangle)$
6. $\quad$ **if** $UP(\Gamma, \delta) \models \Gamma$ **then**
7. $\quad\quad$ **return** SAT
8. $\quad$ **if** $\delta = \langle \rangle$ **then**
9. $\quad\quad$ **return** $UNSAT$
10. $\quad$ **if** *Time to Restart* **then**
11. $\quad\quad \delta \leftarrow \langle \rangle$
12.
13. **end**

6 CDCL with Restarts

We assume the reader is familiar with the standard CDCL-R algorithm. (A self-contained description of CDCL and its relationship to resolution can be found in [16], among other places. The reader may also want to refer to [18] and [2] for distinct presentations of the algorithm, as well as the proofs we work from, and [3] where a careful examination of the relation between resolution and the implication graph method for obtaining conflict clauses appears.)

CDCL-R can be described in terms of a sequence of calls to Probe-and-Learn, as illustrated in Algorithm 2. While many details have been abstracted away, Algorithm 2 captures the core algorithm implemented by CDCL-R-based solvers. For simplicity, let us assume the given formula is unsatisfiable. The algorithm begins with the empty restriction sequence. In the first call to Probe-and-Learn, the restriction sequence is extended until a clause is made false, after which clause learning and back-jumping are carried out (by Handle-Conflict, within Probe-and-Learn). In subsequent executions of the loop body, the restriction sequence resulting from the most recent Handle-Conflict is extended until Probe-and-Learn again finds a conflict. Each asserting clause derived by Handle-Conflict is new, so each call to Probe-and-Learn extends the clause set with at least one new implied clause. This is repeated, until the derived conflict clause is the empty clause.

At this level of abstraction, the signed versions and classical version are not distinguishable, except for the input parameter $\langle T, \prec \rangle$, which affects only the low-level steps in Probe-and-Learn and Handle-Conflict. We make $\langle T, \prec \rangle$ a parameter to make explicit the fact that Probe-and-Learn (and in particular Handle-Conflict), must be appropriate to the order relation and the class of formulas in question. If T is of size 2, then with appropriate choice for Probe-and-Learn, this algorithm is equivalent to the classical CDCL-R.

Since CDCL-R can be viewed simply as a repeated application of Probe-and-Learn, it is straightforward to see that MV-CDCL-R can be guided to refute any formula that has a resolution refutation Π in time polynomial in the size of Π. We need one more kind of object to complete the argument.

Definition 5 (Extended Restriction Sequence for MV-CDCL-R). *An extended restriction sequence for MV-CDCL-R on input Γ and $\langle T, \prec \rangle$ is a finite sequence of symbols satisfying:*

1. *Each symbol is either a literal for $\langle T, \prec \rangle$ or the distinguished symbol R;*

2. *Each maximal sub-sequence with no R is a restriction sequence for Γ.*

We may take two views of an extended restriction sequence. On one view, we may take it as a record or witness of an actual execution of MV-CDCL-R. On the other, we may view it as a string to control an intended execution of MV-CDCL-R.

Lemma 7. *Let Γ be an MV-CNF or Reg-CNF formula of size s over n atoms that has a resolution refutation of size r and width w. Then there is an execution of MV-CDCL-R that refutes Γ in time $O(wn^2 sr)$ and with an underlying resolution refutation of size $O(wn^2 r)$.*

Proof. It is sufficient to show there is an extended restriction sequence that produces such an execution. Since MV-CDCL-R is repeated execution of Probe-and-Learn, we need only to take the sequence of calls to Probe-and-Learn implied by Lemma 6, produce a restriction sequence corresponding to each call, and concatenate all the restriction sequences with an R separating each adjacent pair. □

6.1 Proof Complexity and p-Simulation

Propositional proof complexity is the study of the relative power of proof systems for propositional logic, measured by minimum length of proofs for tautological formulas. The abstract definition of propositional proof system introduced in the seminal paper of Cook and Reckow [8] can be trivially adapted to refutation proofs for unsatisfiable signed CNF formulas (or indeed any co-NP complete set).

Definition 6. *A refutation proof system for a set S of unsatisfiable signed CNF formulas is a set of strings L with a poly-time onto function V_L from strings over the alphabet of L to $S \cup \perp$, such that $V_L(x) = \Gamma$ if x is an L-proof that Γ is unsatisfiable, and $V_L(x) = \perp$ otherwise.*

Intuitively, L is the proofs of the system and V_L is an efficiently computable function that verifies their correctness.

Proof system $\mathcal{A}$ p-simulates proof system $\mathcal{B}$ if there exists a polynomial function $p()$ such that, for every unsatisfiable formula Γ and every $\mathcal{B}$-proof Π_B of Γ, there is an $\mathcal{A}$-proof Π_A of Γ with $|\Pi_A| \leq p(|\Pi_B|)$.

As a simplifying convention, we require that the minimum size of a proof of Γ is $|\Gamma|$. This is not standard in the proof complexity literature, but is necessary for relevance to practical satisfiability algorithms, and is followed also in, e.g., [3, 7, 11, 18]. This is because a formula may be large but have a tiny proof, and any reasonable satisfiability solver begins by reading the entire formula. Moreover, any reasonable CDCL-R-based solver begins by executing unit propagation, which may visit every clause of the formula.

To view a satisfiability algorithm as a proof system, we may take any trace of the algorithm on an unsatisfiable clause set Γ as a proof of the unsatisfiability of Γ, provided that the trace reflects the running time of the algorithm, and that we can efficiently verify that the trace corresponds to an execution of the algorithm that reports "unsatisfiable". For present purposes, we may use extended restriction sequences as MV-CDCL-R proofs.

Theorem 1. *MV-CDCL-R p-simulates MV-RES and Reg-RES.*

Proof. To show that CDCL-R p-simulates resolution, it is sufficient to show that for any resolution refutation Π of clause set Γ there is an extended restriction sequence δ such that, when CDCL-R is executed in accordance with δ on input Γ, it runs in time polynomial in the length of Π, and reports UNSAT. This follows from Lemma 7. □

Corollary 1 (Pipatsrisawat & Darwiche). *CDCL-R p-simulates resolution.*

To see that this is indeed a corollary, it is enough to observe that MV-CNF and MV-RES for $|T| = 2$ are equivalent to the classical case.

Theorem 2. *MV-RES and Reg-RES p-simulate MV-CDCL-R for MV-CNF and Reg-CNF formulas respectively.*

Proof. Consider an execution of MV-CDCL-R that halts and outputs "UNSAT", and let σ be the extended restriction sequence corresponding to this execution. The implicit underlying resolution refutation of Γ consists of the sequence of asserting clauses returned by Handle-Conflict, and all the clauses implicitly used in their derivation. It is clear that this sequence of clauses is of length polynomial in the length of σ. □

7 Application Issues

Our stated motivation was to better understand the question of whether reasoning in multi-valued logics might be effectively carried out by reduction to multi-valued CNF formulas followed by execution of a CDCL-R based algorithm adapted to the multi-valued setting. Our simulation results provide one way of understanding the power of MV-CDCL-R. To the extent that the theorems of [18, 2] are related to positive performance characteristics of solvers based on CDCL-R, we now expect the same to hold for solvers based on MV-CDCL-R. This provides some evidence that reduction to multi-valued CNF may be a fruitful approach to multi-valued logic reasoning.

The MV-CDCL-R algorithm seems quite reasonable to implement. Most of the work is easily inherited from existing CDCL SAT solvers. For $|T|$ up to 64, the sign of a literal can be implemented in a single word on modern CPUs, and key operations such as the check for clashing, computing the residual of two clashing literals, or merging two non-clashing literals, can be done in a single CPU instruction. Even for significantly larger T, most modern CPUs provide support to do these operations in hardware.

A number of algorithms and solvers for signed or multi-valued formulas, or special cases of them, have been described in the literature. Most of these are essentially backtracking or tableau-based. Thus, they correspond to tree-like versions of resolution and we can expect that, as in the classical case, there will be formulas for which they are exponentially less efficient than unrestricted resolution. We are aware of two algorithms that seem closely related to our MV-CDCL-R, namely those of of [14] and [12]. These do not use restarts, which are essential to our work here, but otherwise seem very closely related to the instantiation of MV-CDCL-R for MV-CNF. The interested reader will find a useful discussion of implementation issues in both of those reports. In light of the importance of restarts in high-performance SAT solvers, and of our results here, it would be very interesting to see the effect on performance of modifying these solvers to execute good restart policies.

An alternate reasoning strategy is to reduce the multi-valued reasoning problem to SAT, and then execute a classical (CDCL-R based) SAT solver. The advantage of this approach is that SAT solvers are easily available and subject to constant improvement efforts. We will not address reductions directly from multi-valued logics to SAT, but restrict our attention to the following observation. We can easily reduce Signed CNF satisfiability to SAT as follows. The set of propositional atoms will contain atoms we write as $p = a$, one for each signed atom p and each truth value $a \in T$. Each literal $p \in \{a, b, c, \ldots\}$ then maps to the disjunction $(p = a \vee p = b \vee p = c \vee \ldots)$. Each clause of a Signed CNF formula maps to a single propositional clause. We add to the resulting clauses the set of binary clauses of the form $(\overline{p = a} \vee \overline{p = c})$, for each atom p and each pair $a, c \in T$ with $a \neq c$.

We may simulate Signed resolution using classical resolution on these clauses, but this does not seem of potential use in practice. To simulate the single resolution step

$$\frac{p \in S \vee A \quad p \in R \vee B}{p \in (S \cap R) \vee A \vee B} \quad (3)$$

with classical resolution seems to require a number of 2-valued resolution steps that is quadratic in $|T|$. Since T is fixed, this is only a constant blow-up, but constants can be important in solver design. Moreover, this constant applies to optimal proofs, adding a large amount of non-determinism to the proof simulations. In a practical algorithm, expecting to make $|T|^2$ correct decisions in order to simulate a single multi-valued resolution step seems unreasonably optimistic.

Recently, it was been shown that for any CDCL-R solver S and any unsatisfiable CNF formula F, it is easy to generate a CDCL solver S' (with no restarts) and a formula F' (consisting of a conjunction of F with some new clauses) such that S' generates exactly the same resolution refutation of F that S' does, and with only a small polynomial slow-down [4]. Thus, theoretically, restarts are not really required. However, it is far from clear that this fact can be used to make implemented solvers without restarts that are as fast in practice as solvers that use restarts.

The potential of using resolution-based methods for signed formulas to solve combinatorial optimization problems has been examined in [1]. It would be interesting to relate that direction of work to the present one.

8 Conclusion

We have presented a natural generalization of the SAT algorithm known as CDCL with restarts to signed CNF formulas, in particular to Multi-valued CNF formulas, and to Regular formulas when the truth value set is a lattice. Adapting the proofs from [18, 2] we showed that the algorithm p-simulates natural forms of binary resolution for these formulas, and vice versa. The simulation of resolution by the algorithm is quite efficient, both in terms of length and width. Explaining the impressive performance of SAT solvers in practice in light of their worst-case performance and the NP-completeness of SAT is an area of active interest. We do not know if the theorems of [18, 2] are significant in such an explanation, but it is plausible and consistent with empirical observation. Moreover, there is a good deal of evidence that frequent restarts are important in practice. To the extent that this is so, we have shown that similar properties hold for the generalization to multi-valued CNF formulas. We consider the algorithm we have described easily implementable and of potential use in developing practical model finders and theorem provers for multi-valued logics.

References

[1] Carlos Ansótegui, María Luisa Bonet, Jordi Levy, and Felip Manyà. Resolution procedures for multiple-valued optimization. *Information Sciences*, 227:43 – 59, 2013.

[2] Albert Atserias, Johannes Klaus Fichte, and Marc Thurley. Clause-learning algorithms with many restarts and bounded-width resolution. *J. Artif. Intell. Res. (JAIR)*, 40:353-373, 2011.

[3] Paul Beame, Henry Kautz, and Ashish Sabharwal. Towards understanding and harnessing the potential of clause learning. *Journal of Artificial Intelligence Research*, 22:319-351, 2004.

[4] Paul Beame and Ashish Sabharwal. Non-restarting SAT solvers with simple preprocessing can efficiently simulate resolution. In Carla E. Brodley and Peter Stone, editors, *Proceedings of the Twenty-Eighth AAAI Conference on Artificial Intelligence, July 27 -31, 2014, Québec City, Québec, Canada.*, pages 2608–2615. AAAI Press, 2014.

[5] Bernhard Beckert, Reiner Hähnle, and Felip Manyà. The 2-sat problem of regular signed cnf formulas. In *Proc. 30th IEEE International Symposium on Multi-Valued Logic (ISMVL 2000)*, pages 331–336, 2000.

[6] Bernhard Beckert, Reiner Hahnle, and Filip Manya. The SAT problem of signed CNF formulas. In Basin, D'Agostino, Gabbay, Matthews, and Viganò, editors, *Labelled Deduction*, volume 17 of *Applied Logic Series*, chapter 3, pages 59–80. Kluwer Academic, 2000.

[7] Samuel Buss, Jan Hoffmann, and Jan Johannsen. Resolution trees with lemmas: Resolution refinements that characterize DLL algorithms with clause learning. *Logical Methods in Computer Science*, 4(4:13), 2008.

[8] Stephen Cook and Robert Reckhow. The relative efficiency of propositional proof systems. *J. Symbolic Logic*, 44(1):23–46, 1979.

[9] Reiner Hähnle. Short conjunctive normal forms in finitely valued logics. *J. Log. Comput.*, 4(6):905–927, 1994.

[10] Reiner Hähnle. Exploiting data dependencies in many-valued logics. *Journal of Applied Non-Classical Logics*, 6(1):49–69, 1996.

[11] Philipp Hertel, Fahiem Bacchus, Toniann Pitassi, and Allen Van Gelder. Clause learning can effectively p-simulate general propositional resolution. In *Proc., 23rd National*

Conference on Artificial Intelligence (AAAI-08), Volume 1, pages 283–290, 2008.

[12] Siddhartha Jain, Eoin O'Mahony, and Meinolf Sellmann. A complete multi-valued SAT solver. In *Principles and Practice of Constraint Programming – 16th International Conference (CP 2010), St. Andrews, Scotland, UK, September 6-10, 2010. Proceedings*, pages 281–296, 2010.

[13] Henry Kautz, Eric Horvitz, Yongshao Ruan, Carla Gomes, and Bart Selman. Dynamic restart policies. In *Proc., 19th National Conference on Artificial Intelligence (AAAI-2002)*, pages 674–681, 2002.

[14] Cong Liu, Andreas Kuehlmann, and Matthew W. Moskewicz. Cama: A multi-valued satisfiability solver. In *Proc., 2003 Int'l. Conf. on Computer Aided Design (ICCAD-2003)*, pages 326–333, 2003.

[15] João Marques-Silva and Karem Sakallah. Grasp: A search algorithm for propositional satisfiability. *IEEE Transactions on Computers*, 48(5):506–521, May 1999.

[16] David Mitchell. A SAT solver primer. *EATCS Bulletin*, 85:112–133, February 2005. The Logic in Computer Science Column.

[17] David Mitchell. Resolution and clause learning for multi-valued CNF formulas. In Thomas Lukasiewicz, Rafael Peñaloza, and Anni-Yasmin Turhan, editors, *Proceedings of the First Workshop on Logics for Reasoning about Preferences, Uncertainty, and Vagueness, PRUV 2014, co-located with 7th International Joint Conference on Automated Reasoning (IJCAR 2014), Vienna, Austria, July 23-24, 2014.*, volume 1205 of *CEUR Workshop Proceedings*, pages 141–154. CEUR-WS.org, 2014.

[18] Knot Pipatsrisawat and Adnan Darwiche. On the power of clause-learning SAT solvers as resolution engines. *Artificial Intelligence*, 175(2):512–525, 2011.

[19] Hantao Zhang and Mark E. Stickel. An efficient algorithm for unit propagation. In *Proceedings of the Fourth Int'l Symposium on Artificial Intelligence and Mathematis (AI-MATH'96), Fort Lauderdale Florida USA*, pages 166–169, 1996.

A Linear-time Unit Propagation

We will show that unit propagation can be executed in linear time on MV-CNF and REG-CNF formulas. The method uses a single watched literal in each clause, which is a simplification of the method with head and tail watches of [19].

Lemma 8. *Let Γ be a set of multi-valued clauses with s literals in total, over a domain T of size t. Then $\mathrm{UP}(\Gamma)$, or $\mathrm{UP}(\Gamma,\delta)$, can be computed in time $O(st)$.*

Proof. We assume a data structure for clauses that supports viewing each clause as a list of literals, and each literal $p \sqsubseteq S$ in a clause as a list of atom-value pairs $\langle p, a \rangle$, one for each $a \in S$, in no particular order. (This exact representation is not important, but makes explanation simple.) We further assume that accessing the next literal in a clause or the next value in S takes constant time. For each clause

C we maintain a pointer or index to a single pair $\langle p, a \rangle$ for a single literal $p \mathrel{\mathord{\in}} S$, and call this the "watched value" for C. Initially, the watched value for each clause is the first value of the first literal. As we execute the procedure, we keep track of a set of assignments of values to each atom that remain possible. We will maintain the following invariant:

> If $\langle p, a \rangle$ in the representation of literal $p \mathrel{\mathord{\in}} S$ is the watched value for clause C, then every $\langle p, b \rangle$ that precedes $\langle p, a \rangle$ in the representation of $p \mathrel{\mathord{\in}} S$ is known to be impossible, and for every literal that precedes $p \mathrel{\mathord{\in}} S$ in the representation of C, every value is known to be impossible.

We maintain a queue Q of pairs $\langle p, a \rangle$ of atoms and values which we have determined to be impossible, but for which we have not propagated the effects of that fact. We initialize Q by inserting the set of pairs

$$\{\langle p, a \rangle \mid C = (p \mathrel{\mathord{\in}} S) \text{ is a unit clause of } \Gamma \text{ and } a \notin S\}$$

For each pair $\langle p, a \rangle$ consisting of an atom p and value a, we construct a list of watched occurrences of the pair. Now, until Q is empty, remove one pair $\langle p, a \rangle$ from Q and handle it as follows. Traverse the list of watched occurrences of $\langle p, a \rangle$. For each occurrence, scan the remaining values in the literal with $\langle p, a \rangle$ looking for a value that is not known to be impossible, to be used as a new watch. If none is found, search for one in the subsequent literals of the clause. If none is found, this clause is now effectively an empty clause, so we are done with it and proceed to the next watched occurrence of $\langle p, a \rangle$. If one was found but it is in the last literal of the clause, then this clause is effectively unit, say $(q \mathrel{\mathord{\in}} R)$. Add the set of pairs $\{\langle q, b \rangle \mid b \notin R \text{ and } b \text{ is not already known to be impossible}\}$ to Q. If a new value to watch was found and it is not in the last literal of the clause, then add the corresponding pair to the appropriate watch list, and get the next pair from Q.

This algorithm visits each literal l in Γ at most once, and uses constant work for each such visit, so runs in time $O(st)$. □

Since t is a constant in our analyses, this establishes the linear time unit propagation we need. This algorithm works for both MV-CNF and REG-CNF, and essentially amounts to reducing them to the two-valued case and performing standard watched-literal unit propagation.

Fuzzy Propositional Formulas under the Stable Model Semantics

Joohyung Lee
School of Computing, Informatics, and Decision Systems Engineering
Arizona State University, Tempe, USA
joolee@asu.edu

Yi Wang
School of Computing, Informatics, and Decision Systems Engineering
Arizona State University, Tempe, USA
ywang485@asu.edu

Abstract

We define a stable model semantics for fuzzy propositional formulas, which generalizes both fuzzy propositional logic and the stable model semantics of classical propositional formulas. The syntax of the language is the same as the syntax of fuzzy propositional logic, but its semantics distinguishes stable models from non-stable models. The generality of the language allows for highly configurable nonmonotonic reasoning for dynamic domains involving graded truth degrees. We show that several properties of Boolean stable models are naturally extended to this many-valued setting, and discuss how it is related to other approaches to combining fuzzy logic and the stable model semantics.

1 Introduction

Answer Set Programming (ASP) [18] is a widely applied declarative programming paradigm for the design and implementation of knowledge-intensive applications. One of the attractive features of ASP is its capability to represent the nonmonotonic aspect of knowledge. However, its mathematical basis, the stable model semantics,

---

We are grateful to Joseph Babb, Michael Bartholomew, Enrico Marchioni, and the anonymous referees for their useful comments and discussions related to this paper. We thank Mario Alviano for his help in running the system FASP2SMT. This work was partially supported by the National Science Foundation under Grants IIS-1319794 and IIS-1526301, and by the South Korea IT R&D program MKE/KIAT 2010-TD-300404-001.

is restricted to Boolean values and is too rigid to represent imprecise and vague information. Fuzzy logic, as a form of many-valued logic, can handle such information by interpreting propositions with graded truth degrees, often as real numbers in the interval [0, 1]. The availability of various fuzzy operators gives the user great flexibility in combining the truth degrees. However, the semantics of fuzzy logic is monotonic and is not flexible enough to handle default reasoning as in answer set programming.

Both the stable model semantics and fuzzy logic are generalizations of classical propositional logic in different ways. While they do not subsume each other, it is clear that many real-world problems require both their strengths. This has led to the body of work on combining fuzzy logic and the stable model semantics, known as fuzzy answer set programming (FASP) (e.g., [4, 5, 12, 19, 20, 23, 27, 28]). However, most works consider simple rule forms and do not allow logical connectives nested arbitrarily as in fuzzy logic. An exception is fuzzy equilibrium logic [25], which applies to arbitrary propositional formulas even including strong negation. However, its definition is highly complex.

Unlike the existing works on fuzzy answer set semantics, in this paper, we extend the general stable model semantics from [9, 10] to fuzzy propositional formulas. The syntax of the language is the same as the syntax of fuzzy propositional logic. The semantics, on the other hand, defines a condition which distinguishes *stable* models from non-stable models. The language is a proper generalization of both fuzzy propositional logic and classical propositional formulas under the stable model semantics, and turns out to be essentially equivalent to fuzzy equilibrium logic, but is much simpler. Unlike the interval-based semantics in fuzzy equilibrium logic, the proposed semantics is based on the notions of a reduct and a minimal model, familiar from the usual way stable models are defined, and thus provides a simpler, alternative characterization of fuzzy equilibrium logic. In fact, in the absence of strong negation, a fuzzy equilibrium model always assigns an interval of the form $[v, 1]$ to each atom, which can be simply identified with a single value v under our stable model semantics. Further, we show that strong negation can be eliminated from a formula in favor of new atoms, extending the well-known result in answer set programming. So our simple semantics fully captures fuzzy equilibrium logic.

Also, there is a significant body of work based on the general stable model semantics, such as the splitting theorem [8], the theorem on loop formulas [7], and the concept of aggregates [15]. The simplicity of our semantics would allow for easily extending those results to the many-valued setting, as can be seen from some examples in this paper.

Another contribution of this paper is to show how reasoning about dynamic systems in ASP can be extended to fuzzy ASP. It is well known that actions and their effects on the states of the world can be conveniently represented by answer set programs [16, 17]. On the contrary, to the best of our knowledge, the work on fuzzy ASP has not addressed how the result can be extended to many-valued setting, and most applications discussed so far are limited to static domains only. As a motivating example, consider modeling dynamics of *trust* in social networks. People trust each other in different degrees under some normal assumptions. If person A trusts person B, then A tends to trust person C whom B trusts, to a degree which is positively correlated to the degree to which A trusts B and the degree to which B trusts C. By default, the trust degrees would not change, but may decrease when some conflict arises between the people. Modeling such a domain requires expressing defaults involving fuzzy truth values. We demonstrate how the proposed language can achieve this by taking advantage of its generality over the existing approaches to fuzzy ASP. Thus the generalization is not simply a pure theoretical pursuit but has practical uses in the convenient modeling of defaults involving fuzzy truth values in dynamic domains.

The paper is organized as follows. Section 2 reviews the syntax and the semantics of fuzzy propositional logic, as well as the stable model semantics of classical propositional formulas. Section 3 presents the main definition of a stable model of a fuzzy propositional formula along with several examples. Section 4 presents a generalized definition based on partial degrees of satisfaction and its reduction to the special case, as well as an alternative, second-order logic style definition. Section 5 tells us how the Boolean stable model semantics can be viewed as a special case of the fuzzy stable model semantics, and Section 6 formally compares the fuzzy stable model semantics with normal FASP programs and fuzzy equilibrium logic. Section 7 shows that some well-known properties of the Boolean stable model semantics can be naturally extended to our fuzzy stable model semantics. Section 8 discusses other related work, followed by the conclusion in Section 9. The complete proofs are given in the appendix.

This paper is a significantly extended version of the papers [13, 14]. Instead of the second-order logic style definition used there, we present a new, reduct-based definition as the main definition, which is simpler and more aligned with the standard definition of a stable model. Further, this paper shows that a generalization of the stable model semantics that allows partial degrees of satisfaction can be reduced to the version that allows only the two-valued concept of satisfaction.

2 Preliminaries

2.1 Review: Fuzzy Logic

A fuzzy propositional signature σ is a set of symbols called *fuzzy atoms*. In addition, we assume the presence of a set CONJ of fuzzy conjunction symbols, a set DISJ of fuzzy disjunction symbols, a set NEG of fuzzy negation symbols, and a set IMPL of fuzzy implication symbols.

A *fuzzy (propositional) formula* (of σ) is defined recursively as follows.

- every fuzzy atom $p \in \sigma$ is a fuzzy formula;

- every numeric constant c, where c is a real number in $[0,1]$, is a fuzzy formula;

- if F is a fuzzy formula, then $\neg F$ is a fuzzy formula, where $\neg \in$ NEG;

- if F and G are fuzzy formulas, then $F \otimes G$, $F \oplus G$, and $F \to G$ are fuzzy formulas, where $\otimes \in$ CONJ, $\oplus \in$ DISJ, and $\to \in$ IMPL.

The models of a fuzzy formula are defined as follows [11]. The *fuzzy truth values* are the real numbers in the range $[0,1]$. A *fuzzy interpretation* I of σ is a mapping from σ into $[0,1]$.

The fuzzy operators are functions mapping one or a pair of fuzzy truth values into a fuzzy truth value. Among the operators, $\neg$ denotes a function from $[0,1]$ into $[0,1]$; $\otimes$, $\oplus$, and $\to$ denote functions from $[0,1] \times [0,1]$ into $[0,1]$. The actual mapping performed by each operator can be defined in many different ways, but all of them satisfy the following conditions, which imply that the operators are generalizations of the corresponding classical propositional connectives:[1]

- a fuzzy negation $\neg$ is decreasing, and satisfies $\neg(0) = 1$ and $\neg(1) = 0$;

- a fuzzy conjunction $\otimes$ is increasing, commutative, associative, and $\otimes(1,x) = x$ for all $x \in [0,1]$;

- a fuzzy disjunction $\oplus$ is increasing, commutative, associative, and $\oplus(0,x) = x$ for all $x \in [0,1]$;

- a fuzzy implication $\to$ is decreasing in its first argument and increasing in its second argument; and $\to(1,x) = x$ and $\to(0,0) = 1$ for all $x \in [0,1]$.

Figure 1 lists some specific fuzzy operators that we use in this paper.

---

[1] We say that a function f of arity n is *increasing in its i-th argument* $(1 \le i \le n)$ if $f(arg_1, \ldots, arg_i, \ldots, arg_n) \le f(arg_1, \ldots, arg'_i, \ldots, arg_n)$ whenever $arg_i \le arg'_i$; f is said to be *increasing* if it is increasing in all its arguments. The definition of *decreasing* is similarly defined.

Symbol	Name	Definition
$\otimes_l$	Łukasiewicz t-norm	$\otimes_l(x,y) = max(x+y-1, 0)$
$\oplus_l$	Łukasiewicz t-conorm	$\oplus_l(x,y) = min(x+y, 1)$
$\otimes_m$	Gödel t-norm	$\otimes_m(x,y) = min(x,y)$
$\oplus_m$	Gödel t-conorm	$\oplus_m(x,y) = max(x,y)$
$\otimes_p$	product t-norm	$\otimes_p(x,y) = x \cdot y$
$\oplus_p$	product t-conorm	$\oplus_p(x,y) = x + y - x \cdot y$
$\neg_s$	standard negator	$\neg_s(x) = 1 - x$
$\to_r$	R-implicator induced by $\otimes_m$	$\to_r(x,y) = \begin{cases} 1 & \text{if } x \leq y \\ y & \text{otherwise} \end{cases}$
$\to_s$	S-implicator induced by $\otimes_m$	$\to_s(x,y) = max(1-x, y)$
$\to_l$	Implicator induced by $\otimes_l$	$\to_l(x,y) = min(1-x+y, 1)$

Figure 1: Some t-norms, t-conorms, negator, and implicators

The *truth value* of a fuzzy propositional formula F under I, denoted $v_I(F)$, is defined recursively as follows:

- for any atom $p \in \sigma$, $v_I(p) = I(p)$;

- for any numeric constant c, $v_I(c) = c$;

- $v_I(\neg F) = \neg(v_I(F))$;

- $v_I(F \odot G) = \odot(v_I(F), v_I(G))$ $(\odot \in \{\otimes, \oplus, \to\})$.

(For simplicity, we identify the symbols for the fuzzy operators with the truth value functions represented by them.)

Definition 1. *We say that a fuzzy interpretation I satisfies a fuzzy formula F if $v_I(F) = 1$, and denote it by $I \models F$. We call such I a fuzzy model of F.*

We say that two formulas F and G are equivalent if $v_I(F) = v_I(G)$ for all interpretations I, and denote it by $F \Leftrightarrow G$.

2.2 Review: Stable Models of Classical Propositional Formulas

A *propositional signature* is a set of symbols called *atoms*. A propositional formula is defined recursively using atoms and the following set of primitive propositional connectives: $\bot$, $\top$, $\neg$, $\wedge$, $\vee$, $\to$.

An *interpretation* of a propositional signature is a function from the signature into {FALSE, TRUE}. We identify an interpretation with the set of atoms that are true in it.

A *model* of a propositional formula is an interpretation that *satisfies* the formula. According to [10], the models are divided into stable models and non-stable models as follows. The *reduct* F^X of a propositional formula F relative to a set X of atoms is the formula obtained from F by replacing every maximal subformula that is not satisfied by X with $\bot$. Set X is called a *stable model* of F if X is a minimal set of atoms satisfying F^X.

Alternatively, the reduct F^X can be defined recursively as follows:

Definition 2.
- When F is an atom or $\bot$ or $\top$, $F^X = \begin{cases} F & \text{if } X \models F; \\ \bot & \text{otherwise;} \end{cases}$

- $(\neg F)^X = \begin{cases} \bot & \text{if } X \models F; \\ \top & \text{otherwise;} \end{cases}$

- For $\odot \in \{\land, \lor, \to\}$,
$$(F \odot G)^X = \begin{cases} F^X \odot G^X & \text{if } X \models F \odot G; \\ \bot & \text{otherwise.} \end{cases}$$

In the next section, we extend this definition to cover fuzzy propositional formulas.

3 Definition and Examples

3.1 Reduct-based Definition

Let σ be a fuzzy propositional signature, F a fuzzy propositional formula of σ, and I an interpretation of σ.

Definition 3. *The (fuzzy) reduct of F relative to I, denoted F^I, is defined recursively as follows:*

- *For any fuzzy atom or numeric constant F, $F^I = F$;*

- $(\neg F)^I = v_I(\neg F);$

- $(F \odot G)^I = (F^I \odot G^I) \otimes_m v_I(F \odot G)$, where $\odot \in \{\otimes, \oplus, \to\}$.

Compare this definition and Definition 2. They are structurally similar, but also have subtle differences. One of them is that in the case of binary connectives $\odot$, instead of distinguishing the cases between when X satisfies the formula or not as in Definition 2, Definition 3 keeps a conjunction of $F^I \odot G^I$ and $v_I(F \odot G)$.[2]

Another difference is that in the case of an atom, Definition 3 is a bit simpler as it does not distinguish between the two cases. It turns out that the same clause can be applied to Definition 2 (i.e., $F^X = F$ when F is an atom regardless of $X \models F$), still yielding an equivalent definition of a Boolean stable model. So this difference is not essential.

For any two fuzzy interpretations J and I of signature σ and any subset $\mathbf{p}$ of σ, we write $J \leq^{\mathbf{p}} I$ if

- $J(p) = I(p)$ for each fuzzy atom not in $\mathbf{p}$, and

- $J(p) \leq I(p)$ for each fuzzy atom p in $\mathbf{p}$.

We write $J <^{\mathbf{p}} I$ if $J \leq^{\mathbf{p}} I$ and $J \neq I$. We may simply write it as $J < I$ when $\mathbf{p} = \sigma$.

Definition 4. *We say that an interpretation I is a (fuzzy) stable model of F relative to $\mathbf{p}$ (denoted $I \models \mathrm{SM}[F; \mathbf{p}]$) if*

- *$I \models F$, and*

- *there is no interpretation J such that $J <^{\mathbf{p}} I$ and J satisfies F^I.*

If $\mathbf{p}$ is the same as the underlying signature, we may simply write $\mathrm{SM}[F; \mathbf{p}]$ as $\mathrm{SM}[F]$ and drop the clause "relative to $\mathbf{p}$."

We call an interpretation J such that $J <^{\mathbf{p}} I$ and J satisfies F^I as a witness to dispute the stability of I (for F relative to $\mathbf{p}$). In other words, a model of F is stable if it has no witness to dispute its stability for F.

Clearly, when $\mathbf{p}$ is empty, Definition 4 reduces to the definition of a fuzzy model in Definition 1 simply because there is no interpretation J such that $J <^{\emptyset} I$.

The definition of a reduct can be simplified in the cases of $\otimes$ and $\oplus$, which are increasing in both arguments:

- $(F \otimes G)^I = (F^I \otimes G^I); \quad (F \oplus G)^I = (F^I \oplus G^I).$

This is due to the following proposition, which can be proved by induction.

---

[2] In fact, a straightforward modification of the second subcases in Definition 2 by replacing $\bot$ with some fixed truth values does not work for the fuzzy stable model semantics.

Proposition 1. *For any interpretations I and J such that $J \leq^{\mathbf{p}} I$, it holds that*
$$v_J(F^I) \leq v_I(F).$$

In the following, we will assume this simplified form of a reduct.

Also, we may view $\neg F$ as shorthand for some fuzzy implication $F \to 0$. For instance, what we called the standard negator can be derived from the residual implicator $\to_l$ induced by Łukasiewicz t-norm, defined as $\to_l(x,y) = min(1-x+y, 1)$. In view of Proposition 1, for $J \leq^{\mathbf{p}} I$,

$$v_J((F \to_l 0)^I) = v_J((F^I \to_l 0) \otimes_m v_I(F \to_l 0))$$
$$= v_J(v_I(F \to_l 0)) = v_I(F \to_l 0) = v_I(\neg_s F).$$

Thus the second clause of Definition 3 can be viewed as a special case of the third clause.

Example 1. *Consider the fuzzy formula $F_1 = \neg_s q \to_r p$ and the interpretation $I = \{(p, x), (q, 1-x)\}$, where $x \in [0,1]$. $I \models F_1$ because*

$$v_I(\neg_s q \to_r p) = \to_r(x, x) = 1.$$

The reduct F_1^I is

$$((\neg_s q)^I \to_r p) \otimes_m 1 \Leftrightarrow (v_I(\neg_s q) \to_r p) \Leftrightarrow x \to_r p$$

I can be a stable model only if $x = 1$. Otherwise, $\{(p, x), (q, 0)\}$ is a witness to dispute the stability of I.

On the other hand, for $F_2 = (\neg_s q \to_r p) \otimes_m (\neg_s p \to_r q)$, for any value of $x \in [0,1]$, I is a stable model. [3]

$$\begin{aligned} F_2^I &= (\neg_s q \to_r p)^I &\otimes_m (\neg_s p \to_r q)^I \\ &\Leftrightarrow (v_I(\neg_s q) \to_r p) \otimes_m 1 &\otimes_m (v_I(\neg_s p) \to_r q) \otimes_m 1 \\ &\Leftrightarrow (x \to_r p) &\otimes_m ((1-x) \to_r q). \end{aligned}$$

No interpretation J such that $J <^{\{p,q\}} I$ satisfies F_2^I.

Example 2.
- $F_1 = p \to_r p$ is a tautology (i.e., every interpretation is a model of the formula), but not all models are stable. First, $I_1 = \{(p, 0)\}$ is a stable model. The reduct $F_1^{I_1}$ is

$$(p^I \to_r p^I) \otimes_m 1 \Leftrightarrow p \to_r p \Leftrightarrow 1$$

---

[3] Strictly speaking, $(1-x)$ in the reduct should be understood as the value of the arithmetic function applied to the arguments.

and obviously, there is no witness to dispute the stability of I.

No other interpretation $I_2 = \{(p,x)\}$ where $x > 0$ is a stable model. The reduct $F_1^{I_2}$ is again equivalent to 1, but I_1 serves as a witness to dispute the stability of I_2.

- $F_2 = \neg_s \neg_s p \to_r p$ is equivalent to F_1, but their stable models are different. Any $I = \{(p,x)\}$, where $x \in [0,1]$, is a stable model of F_2. The reduct F_2^I is

$$((\neg_s \neg_s p)^I \to_r p^I) \otimes_m 1 \;\Leftrightarrow\; v_I(\neg_s \neg_s p) \to_r p$$
$$\Leftrightarrow\; x \to_r p.$$

No interpretation J such that $J <^{\{p\}} I$ satisfies F_2^I.

- Let F_3 be $\neg_s p \oplus_l p$. Any $I = \{(p,x)\}$, where $x \in [0,1]$, is a stable model of F_3. The reduct F_3^I is

$$(\neg_s p)^I \oplus_l p^I \;\Leftrightarrow\; v_I(\neg_s p) \oplus_l p \;\Leftrightarrow\; (1-x) \oplus_l p.$$

No interpretation J such that $J <^p I$ satisfies the reduct.

The following proposition extends a well-known fact about the relationship between a formula and its reduct in terms of satisfaction.

Proposition 2. *A (fuzzy) interpretation I satisfies a formula F if and only if I satisfies F^I.*

For any fuzzy formula F, any interpretation I and any set $\mathbf{p}$ of atoms, we say that I is a *minimal* model of F relative to $\mathbf{p}$ if I satisfies F and there is no interpretation J such that $J <^{\mathbf{p}} I$ and J satisfies F. Using this notion, Proposition 2 tells us that Definition 4 can be reformulated as follows.

Corollary 1. *An interpretation I is a (fuzzy) stable model of F relative to $\mathbf{p}$ iff I is a minimal model of F^I relative to $\mathbf{p}$.*

This reformulation relies on the fact that $\otimes_m$ is intended in the third bullet of Definition 3 instead of an arbitrary fuzzy conjunction because we want the truth value of the "conjunction" of $F^I \to G^I$ and $F \to G$ to be either the truth value of $F^I \to G^I$ or the truth value $F \to G$. Conjunctions that do not have this property lead to unintuitive behaviors, such as violating Proposition 2. As an example, consider the formula

$$F = 0.6 \to_r (1 \to_r p)$$

and the interpretation
$$I = \{(p, 0.6)\}.$$
Clearly, I satisfies F. According to Definition 3,
$$\begin{aligned}
F^I &= (0.6 \to_r (1 \to_r p)^I) \otimes_m v_I(0.6 \to_r (1 \to_r p)) \\
&= (0.6 \to_r ((1 \to_r p) \otimes_m v_I(1 \to_r p))) \otimes_m 1 \\
&= 0.6 \to_r ((1 \to_r p) \otimes_m 0.6)
\end{aligned}$$

and
$$\begin{aligned}
v_I(F^I) &= (0.6 \to_r ((1 \to_r 0.6) \otimes_m 0.6)) \\
&= 0.6 \to_r 0.6 \ = \ 1
\end{aligned}$$

So I satisfies F^I as well. However, if we replace $\otimes_m$ by $\otimes_l$ in the third bullet of Definition 3, we get
$$\begin{aligned}
v_I(F^I) &= 0.6 \to_r ((1 \to_r 0.6) \otimes_l 0.6) \\
&= 0.6 \to_r 0.2 \ = \ 0.2 \ < \ 1,
\end{aligned}$$

which indicates that I does not satisfy F^I. Therefore, I is a fuzzy stable model of F according to Definition 4, but it is not even a model of the reduct F^I if $\otimes_l$ were used in place of $\otimes_m$ in the definition of a reduct.

Similarly, it can be checked that the same issue remains if we use $\otimes_p$ in place of $\otimes_m$.

The following example illustrates how the commonsense law of inertia involving fuzzy truth values can be represented.

Example 3. Let σ be $\{p_0, np_0, p_1, np_1\}$ and let F be $F_1 \otimes_m F_2$, where F_1 represents that p_1 and np_1 are complementary, i.e., the sum of their truth values is 1: [4]
$$F_1 = \neg_s(p_1 \otimes_l np_1) \otimes_m \neg_s\neg_s(p_1 \oplus_l np_1).$$

---

[4] This is similar to the formulas used under the Boolean stable model semantics to express that two Boolean atoms p_1 and np_1 take complimentary values, i.e.,
$$\neg(p_1 \wedge np_1) \wedge \neg\neg(p_1 \vee np_1).$$

F_2 represents that by default p_1 has the truth value of p_0, and np_1 has the truth value of np_0:[5]

$$F_2 = ((p_0 \otimes_m \neg_s \neg_s p_1) \to_r p_1) \otimes_m ((np_0 \otimes_m \neg_s \neg_s np_1) \to_r np_1).$$

One can check that the interpretation

$$I_1 = \{(p_0, x), (np_0, 1-x), (p_1, x), (np_1, 1-x)\}$$

(x is any value in $[0,1]$) is a stable model of F relative to $\{p_1, np_1\}$: F^{I_1} is equivalent to

$$(p_0 \otimes_m x \to_r p_1) \otimes_m (np_0 \otimes_m (1-x) \to_r np_1).$$

No interpretation J such that $J <^{\{p_1, np_1\}} I_1$ satisfies F^{I_1}.

The interpretation

$$I_2 = \{(p_0, x), (np_0, 1-x), (p_1, y), (np_1, 1-y)\},$$

where $y > x$, is not a stable model of F. The reduct F^{I_2} is equivalent to

$$(p_0 \otimes_m y \to_r p_1) \otimes_m (np_0 \otimes_m (1-y) \to_r np_1),$$

and the interpretation $\{(p_0, x), (np_0, 1-x), (p_1, x), (np_1, 1-y)\}$ serves as a witness to dispute the stability of I_2.

Similarly, when $y < x$, we can check that I_2 is not a stable model of F relative to $\{p_1, np_1\}$.

On the other hand, if we conjoin F with $y \to_r p_1$ and $(1-y) \to_r np_1$ to yield $F \otimes_m (y \to_r p_1) \otimes_m ((1-y) \to_r np_1)$, then the default behavior is overridden, and I_2 is a stable model of $F \otimes_m (y \to_r p_1) \otimes_m ((1-y) \to_r np_1)$ relative to $\{p_1, np_1\}$.[6]

This behavior is useful in expressing the commonsense law of inertia involving fuzzy values. Suppose p_0 represents the value of fluent p at time 0, and p_1 represents the value at time 1. Then F states that "by default, the fluent retains the previous value." The default value is overridden if there is an action that sets p to a different value.

This way of representing the commonsense law of inertia is a straightforward extension of the solution in ASP.

---

[5] This is similar to the rules used in ASP to express the commonsense law of inertia, e.g.,

$$p_0 \wedge \neg\neg p_1 \to p_1.$$

[6] One may wonder why the part $(1-y) \to_r np_1$ is also needed. It can be checked that if we drop the part $(1-y) \to_r np_1$ and have y less than x, then I_2 (with $y < x$) is not a stable model of $F \otimes_m (y \to_r p_1)$ relative to $\{p_1, np_1\}$ because $J = \{(p_0, x), (np_0, 1-x), (p_1, y), (np_1, 1-x)\}$ disputes the stability of I_2.

Example 4. *The trust example in the introduction can be formalized in the fuzzy stable model semantics as follows. Below x, y, z are schematic variables ranging over people, and t is a schematic variable ranging over time steps. $Trust(x,y,t)$ is a fuzzy atom representing that "x trusts y at time t." Similarly, $Distrust(x,y,t)$ is a fuzzy atom representing that "x distrusts y at time t."*

The trust relation is reflexive:

$$F_1 = Trust(x,x,t).$$

The trust and distrust degrees are complementary, i.e., their sum is 1 (similar to F_1 in Example 3):

$$F_2 = \neg_s(Trust(x,y,t) \otimes_l Distrust(x,y,t)),$$
$$F_3 = \neg_s\neg_s(Trust(x,y,t) \oplus_l Distrust(x,y,t)).$$

Initially, if x trusts y to degree d_1 and y trusts z to degree d_2, then we assume x trusts z to degree $d_1 \times d_2$; furthermore the initial distrust degree is 1 minus the initial trust degree.

$$F_4 = Trust(x,y,0) \otimes_p Trust(y,z,0) \to_r Trust(x,z,0),$$
$$F_5 = \neg_s Trust(x,y,0) \to_r Distrust(x,y,0).$$

The inertia assumption (similar to F_2 in Example 3):

$$F_6 = Trust(x,y,t) \otimes_m \neg_s\neg_s Trust(x,y,t+1) \to_r Trust(x,y,t+1),$$
$$F_7 = Distrust(x,y,t) \otimes_m \neg_s\neg_s Distrust(x,y,t+1) \to_r Distrust(x,y,t+1).$$

A conflict increases the distrust degree by the conflict degree:

$$F_8 = Conflict(x,y,t) \oplus_l Distrust(x,y,t) \to_r Distrust(x,y,t+1),$$

Let F_{TW} be $F_1 \otimes_m F_2 \otimes_m \cdots \otimes_m F_8$. Suppose we have the formula $F_{Fact} = Fact_1 \otimes_m Fact_2$ that gives the initial trust degree.

$$Fact_1 = 0.8 \to_r Trust(Alice, Bob, 0),$$
$$Fact_2 = 0.7 \to_r Trust(Bob, Carol, 0).$$

Although there is no fact about how much Alice trusts Carol, any stable model of $F_{TW} \otimes_m F_{Fact}$ assigns value 0.56 to the atom $Trust(Alice, Carol, 0)$. On the other hand, the stable model assigns value 0 to $Trust(Alice, David, 0)$ due to the closed world assumption under the stable model semantics.

When we conjoin $F_{TW} \otimes F_{Fact}$ with $0.2 \rightarrow_r \text{Conflict}(Alice, Carol, 0)$, the stable model of
$$F_{TW} \otimes_m F_{Fact} \otimes_m (0.2 \rightarrow_r \text{Conflict}(Alice, Carol, 0))$$
manifests that the trust degree between Alice and Carol decreases to 0.36 at time 1. More generally, if we have more actions that change the trust degree in various ways, by specifying the entire history of actions, we can determine the evolution of the trust distribution among all the participants. Useful decisions can be made based on this information. For example, Alice may decide not to share her personal pictures to those whom she trusts less than degree 0.48.

Note that this example, like Example 3, uses nested connectives, such as $\neg_s\neg_s$, that are not available in the syntax of FASP considered in earlier work, such as [12, 19].

It is often assumed that, for any fuzzy rule arrow $\leftarrow$, it holds that $\leftarrow(x,y) = 1$ iff $x \geq y$ [5]. This condition is required to use $\leftarrow$ to define an immediate consequence operator for a positive program whose least fixpoint coincides with the unique minimal model. (A positive program is a set of rules of the form $a \leftarrow b_1 \otimes \ldots \otimes b_n$, where $a, b_1, \ldots, b_n$ are atoms.) Notice that $\rightarrow_r$ in Figure 1 satisfies this condition, but $\rightarrow_s$ does not.

4 Generalization and Alternative Definitions

4.1 y-Stable Models

While the fuzzy stable model semantics presented in the previous section allows atoms to have many values, like ASP, it holds on to the two-valued concept of satisfaction, i.e., a formula is either satisfied or not. In a more flexible setting we may allow a formula to be partially satisfied to a certain degree.

First, we generalize the notion of satisfaction to allow partial degrees of satisfaction as in [27].

Definition 5. *For any real number $y \in [0,1]$, we say that a fuzzy interpretation I y-satisfies a fuzzy formula F if $v_I(F) \geq y$, and denote it by $I \models_y F$. We call I a fuzzy y-model of F.*

Using this generalized notion of satisfaction, it is straightforward to generalize the definition of a stable model to incorporate partial degrees of satisfaction.

Definition 6. *We say that an interpretation I is a fuzzy y-stable model of F relative to $\mathbf{p}$ (denoted $I \models_y \text{SM}[F; \mathbf{p}]$) if*

- $I \models_y F$ and

- there is no interpretation J such that $J <^{\mathbf{p}} I$ and $J \models_y F^I$.

Fuzzy models and fuzzy stable models as defined in Section 2.1 and Section 3 are fuzzy 1-models and fuzzy 1-stable models according to these generalized definitions.

Example 5. *Consider the fuzzy formula $F = \neg_s p \to_r q$ and the interpretation $I = \{(p,0),(q,0.6)\}$. $I \models_{0.6} F$ because $v_I(\neg_s p \to_r q) = \to_r(1,0.6) = 0.6$. The reduct F^I is*

$$((\neg_s p)^I \to_r q) \otimes_m 1 \Leftrightarrow (v_I(\neg_s p) \to_r q) \Leftrightarrow 1 \to_r q$$

Clearly, for any J such that $J <^{\{p,q\}} I$, we observe that $J \not\models_{0.6} F^I$. Hence, I is a 0.6-stable model of F.

On the other hand, the generalized definition is not essential in the sense that it can be reduced to the special case as follows.

Theorem 1. *For any fuzzy formula F, an interpretation I is a y-stable model of F relative to $\mathbf{p}$ iff I is a 1-stable model of $y \to F$ relative to $\mathbf{p}$ as long as the implication $\to$ satisfies the condition $\to(x,y) = 1$ iff $y \geq x$.*

For example, in accordance with Theorem 1, $\{(p, 0.6)\}$ is a 0.6-stable model of $\neg_s p \to_r q$, as well as a 1-stable model of $0.6 \to_r (\neg_s p \to_r q)$.

4.2 Second-Order Logic Style Definiton

In [9], second-order logic was used to define the stable models of a first-order formula, which is equivalent to the reduct-based definition when the domain is finite. Similarly, but instead of using second-order logic, we can express the same concept using auxiliary atoms that do not belong to the original signature.

Let σ be a set of fuzzy atoms, and let $\mathbf{p} = (p_1, \ldots, p_n)$ be a list of distinct atoms belonging to σ, and let $\mathbf{q} = (q_1, \ldots, q_n)$ be a list of new, distinct fuzzy atoms not belonging to σ. For two interpretations I and J of σ that agree on all atoms in $\sigma \setminus \mathbf{p}$, $I \cup J_{\mathbf{q}}^{\mathbf{p}}$ denotes the interpretation of $\sigma \cup \mathbf{q}$ such that

- $(I \cup J_{\mathbf{q}}^{\mathbf{p}})(p) = I(p)$ for each atom p in σ, and

- $(I \cup J_{\mathbf{q}}^{\mathbf{p}})(q_i) = J(p_i)$ for each $q_i \in \mathbf{q}$.

For any fuzzy formula F of signature σ, $F^*(\mathbf{q})$ is defined as follows.

- $p_i^* = q_i$ for each $p_i \in \mathbf{p}$;

- $F^* = F$ for any atom $F \notin \mathbf{p}$ or any numeric constant F;
- $(\neg F)^* = \neg F$;
- $(F \otimes G)^* = F^* \otimes G^*$; $(F \oplus G)^* = F^* \oplus G^*$;
- $(F \to G)^* = (F^* \to G^*) \otimes_m (F \to G)$.

Theorem 2. *A fuzzy interpretation I is a fuzzy stable model of F relative to $\mathbf{p}$ iff*

- $I \models F$, *and*
- *there is no fuzzy interpretation J such that $J <^{\mathbf{p}} I$ and $I \cup J_{\mathbf{q}}^{\mathbf{p}} \models F^*(\mathbf{q})$.*

5 Relation to Boolean-Valued Stable Models

The Boolean stable model semantics in Section 2.2 can be embedded into the fuzzy stable model semantics as follows:

For any classical propositional formula F, define F^{fuzzy} to be the fuzzy propositional formula obtained from F by replacing $\bot$ with 0, $\top$ with 1, $\neg$ with $\neg_s$, $\wedge$ with $\otimes_m$, $\vee$ with $\oplus_m$, and $\to$ with $\to_s$. We identify the signature of F^{fuzzy} with the signature of F. Also, for any propositional interpretation I, we define the corresponding fuzzy interpretation I^{fuzzy} as

- $I^{fuzzy}(p) = 1$ if $I(p) = \text{TRUE}$;
- $I^{fuzzy}(p) = 0$ otherwise.

The following theorem tells us that the Boolean-valued stable model semantics can be viewed as a special case of the fuzzy stable model semantics.

Theorem 3. *For any classical propositional formula F and any classical propositional interpretation I, I is a stable model of F relative to $\mathbf{p}$ iff I^{fuzzy} is a stable model of F^{fuzzy} relative to $\mathbf{p}$.*

Example 6. *Let F be the classical propositional formula $\neg q \to p$. F has only one stable model $I = \{p\}$. Likewise, as shown in Example 1, $F^{fuzzy} = \neg_s q \to_s p$ has only one fuzzy stable model $I^{fuzzy} = \{(p, 1), (q, 0)\}$.*

Theorem 3 may not necessarily hold for different selections of fuzzy operators, as illustrated by the following example.

Example 7. Let F be the classical propositional formula $p \vee p$. Classical interpretation $I = \{p\}$ is a stable model of F. However, $I^{fuzzy} = \{(p,1)\}$ is not a stable model of $F^{fuzzy} = p \oplus_l p$ because there is $J = \{(p, 0.5)\} < I$ that satisfies $(F^{fuzzy})^I = p \oplus_l p$.

However, one direction of Theorem 3 still holds for different selections of fuzzy operators.

Theorem 4. *For any classical propositional formula F, let F_1^{fuzzy} be the fuzzy formula obtained from F by replacing $\bot$ with 0, $\top$ with 1, $\neg$ with any fuzzy negation symbol, $\wedge$ with any fuzzy conjunction symbol, $\vee$ with any fuzzy disjunction symbol, and $\rightarrow$ with any fuzzy implication symbol. For any classical propositional interpretation I, if I^{fuzzy} is a fuzzy stable model of F_1^{fuzzy} relative to $\mathbf{p}$, then I is a Boolean stable model of F relative to $\mathbf{p}$.*

6 Relation to Other Approaches to Fuzzy ASP

6.1 Relation to Stable Models of Normal FASP Programs

A normal FASP program [19] is a finite set of rules of the form

$$a \leftarrow b_1 \otimes \ldots \otimes b_m \otimes \neg b_{m+1} \otimes \ldots \otimes \neg b_n,$$

where $n \geq m \geq 0$, $a, b_1, \ldots, b_n$ are fuzzy atoms or numeric constants in $[0,1]$, and $\otimes$ is any fuzzy conjunction. We identify the rule with the fuzzy implication

$$b_1 \otimes \ldots \otimes b_m \otimes \neg_s b_{m+1} \otimes \ldots \otimes \neg_s b_n \rightarrow_r a,$$

which allows us to say that a fuzzy interpretation I of signature σ *satisfies* a rule R if $v_I(R) = 1$. I *satisfies* an FASP program Π if I satisfies every rule in Π.

According to [19], an interpretation I is a *fuzzy answer set* of a normal FASP program Π if I satisfies Π, and no interpretation J such that $J <^\sigma I$ satisfies the reduct of Π w.r.t. I, which is the program obtained from Π by replacing each negative literal $\neg b$ with the constant for $1 - I(b)$.

Theorem 5. *For any normal FASP program $\Pi = \{r_1, \ldots, r_n\}$, let F be the fuzzy formula $r_1 \otimes \ldots \otimes r_n$, where $\otimes$ is any fuzzy conjunction. An interpretation I is a fuzzy answer set of Π in the sense of [19] if and only if I is a stable model of F.*

Example 8. *Let Π be the following program*

$$p \leftarrow \neg q$$
$$q \leftarrow \neg p.$$

The answer sets of Π according to [19] are $\{(p,x),(q,1-x)\}$, where x is any value in $[0,1]$: the corresponding fuzzy formula F is $(\neg_s q \to_r p) \otimes_m (\neg_s p \to_r q)$; As we observed in Example 1, its stable models are $\{(p,x),(q,1-x)\}$, where x is any real number in $[0,1]$.

6.2 Relation to Fuzzy Equilibrium Logic

Like the fuzzy stable model semantics introduced in this paper, fuzzy equilibrium logic [25] generalizes fuzzy ASP programs to arbitrary propositional formulas, but its definition is quite complex as it is based on some complex operations on pairs of intervals and considers strong negation as one of the primitive connectives. Nonetheless, we show that fuzzy equilibrium logic is essentially equivalent to the fuzzy stable model semantics where all atoms are subject to minimization.

6.2.1 Review: Fuzzy Equilibrium Logic

We first review the definition of fuzzy equilibrium logic from [25]. The syntax is the same as the one we reviewed in Section 2.1 except that a new connective $\sim$ (strong negation) may appear in front of atoms.[7] For any fuzzy propositional signature σ, a (fuzzy N5) *valuation* is a mapping from $\{h,t\} \times \sigma$ to subintervals of $[0,1]$ such that $V(t,a) \subseteq V(h,a)$ for each atom $a \in \sigma$. For $V(w,a) = [u,v]$, where $w \in \{h,t\}$, we write $V^-(w,a)$ to denote the lower bound u and $V^+(w,a)$ to denote the upper bound v. The *truth value* of a fuzzy formula under V is defined as follows.

- $V(w,c) = [c,c]$ for any numeric constant c;
- $V(w, \sim a) = [1 - V^+(w,a), 1 - V^-(w,a)]$, where $\sim$ is the symbol for strong negation;
- $V(w, F \otimes G) = [V^-(w,F) \otimes V^-(w,G), \ V^+(w,F) \otimes V^+(w,G)]$;[8]
- $V(w, F \oplus G) = [V^-(w,F) \oplus V^-(w,G), \ V^+(w,F) \oplus V^+(w,G)]$;
- $V(h, \neg F) = [1 - V^-(t,F), \ 1 - V^-(h,F)]$;
- $V(t, \neg F) = [1 - V^-(t,F), \ 1 - V^-(t,F)]$;
- $V(h, F \to G) = [min(V^-(h,F) \to V^-(h,G), V^-(t,F) \to V^-(t,G)),$
  $V^-(h,F) \to V^+(h,G)]$;

---

[7]The definition from [25] allows strong negation in front of any formulas. We restrict its occurrence only in front of atoms as usual in answer set programs.
[8]For readability, we write the infix notation $(x \odot y)$ in place of $\odot(x,y)$.

- $V(t, F \to G) = [V^-(t,F) \to V^-(t,G),\ V^-(t,F) \to V^+(t,G)]$.

A valuation V is a (fuzzy N5) model of a formula F if $V^-(h, F) = 1$, which implies $V^+(h, F) = V^-(t, F) = V^+(t, F) = 1$. For two valuations V and V', we say $V' \preceq V$ if $V'(t, a) = V(t, a)$ and $V(h, a) \subseteq V'(h, a)$ for all atoms a. We say $V' \prec V$ if $V' \preceq V$ and $V' \neq V$. We say that a model V of F is *h-minimal* if there is no model V' of F such that $V' \prec V$. An h-minimal fuzzy N5 model V of F is a *fuzzy equilibrium model* of F if $V(h, a) = V(t, a)$ for all atoms a.

6.2.2 In the Absence of Strong Negation

We first establish the correspondence between fuzzy stable models and fuzzy equilibrium models in the absence of strong negation. As in [25], we assume that the fuzzy negation $\neg$ is $\neg_s$.

Notice that a fuzzy equilibrium model assigns an interval of values to each atom, rather than a single value as in fuzzy stable models. This accounts for the complexity of the definition of a fuzzy model. However, it turns out that in the absence of strong negation, all upper bounds assigned by a fuzzy equilibrium model are 1.

Lemma 1. *Given a formula F containing no strong negation, any equilibrium model V of F satisfies $V^+(h, a) = V^+(t, a) = 1$ for all atoms a.*

Therefore, in the absence of strong negation, any equilibrium model can be identified with a fuzzy interpretation as follows. For any valuation V, we define a fuzzy interpretation I_V as $I_V(p) = V^-(h, p)$ for each atom $p \in \sigma$.

The following theorem asserts that there is a 1-1 correspondence between fuzzy equilibrium models and fuzzy stable models.

Theorem 6. *Let F be a fuzzy propositional formula of σ that contains no strong negation.*

(a) *A valuation V of σ is a fuzzy equilibrium model of F iff $V^-(h, p) = V^-(t, p)$, $V^+(h, p) = V^+(t, p) = 1$ for all atoms p in σ and I_V is a stable model of F relative to σ.*

(b) *An interpretation I of σ is a stable model of F relative to σ iff $I = I_V$ for some fuzzy equilibrium model V of F.*

6.2.3 Allowing Strong Negation

In this section, we extend the relationship between fuzzy equilibrium logic and our stable model semantics by allowing strong negation. This is done by simulating strong negation by new atoms in our semantics.

Let σ denote the signature. For a fuzzy formula F over σ that may contain strong negation, define F' over $\sigma \cup \{np \mid p \in \sigma\}$ as the formula obtained from F by replacing every strongly negated atom $\sim p$ with a new atom np. The transformation $nneg(F)$ ("no strong negation") is defined as $nneg(F) = F' \otimes_m \bigotimes_{p \in \sigma} \neg_s(p \otimes_l np)$.

For any valuation V of signature σ, we define the valuation $nneg(V)$ of $\sigma \cup \{np \mid p \in \sigma\}$ as

$$\begin{cases} nneg(V)(w,p) = [V^-(w,p), 1] \\ nneg(V)(w,np) = [1 - V^+(w,p), 1] \end{cases}$$

for all atoms $p \in \sigma$. Clearly, for every valuation V of σ, there exists a corresponding interpretation $\mathsf{I}_{nneg(V)}$ of $\sigma \cup \{np \mid p \in \sigma\}$. On the other hand, there exists an interpretation I of $\sigma \cup \{np \mid p \in \sigma\}$ for which there is no corresponding valuation V of σ such that $I = \mathsf{I}_{nneg(V)}$.

Example 9. *Suppose $\sigma = \{p\}$. For the valuation V such that $V(w,p) = [0.2, 0.7]$, $nneg(V)$ is a valuation of $\{p, np\}$ such that*

$$nneg(V)(w,p) = [0.2, 1] \text{ and } nneg(V)(w,np) = [0.3, 1].$$

Further, $\mathsf{I}_{nneg(V)}$ is an interpretation of $\{p, np\}$ such that

$$\mathsf{I}_{nneg(V)}(p) = 0.2 \text{ and } \mathsf{I}_{nneg(V)}(np) = 0.3.$$

On the other hand, the interpretation $I = \{(p, 0.6), (np, 0.8)\}$ of $\sigma \cup \{np \mid p \in \sigma\}$ has no corresponding valuation V of σ such that $I = \mathsf{I}_{nneg(V)}$ because $[0.6, 0.2]$ is not a valid valuation.

The following proposition asserts that strong negation can be eliminated in favor of new atoms, extending the well-known results with the Boolean stable model semantics [9, Section 8] to fuzzy formulas.

Proposition 3. *For any fuzzy formula F that may contain strong negation, a valuation V is an equilibrium model of F iff $nneg(V)$ is an equilibrium model of $nneg(F)$.*

Proposition 3 allows us to extend the 1-1 correspondence between fuzzy equilibrium models and fuzzy stable models in Theorem 6 to any formula that contains strong negation.

Theorem 7. *For any fuzzy formula F of signature σ that may contain strong negation,*

(a) *A valuation V of σ is a fuzzy equilibrium model of F iff $V(h,p) = V(t,p)$ for all atoms p in σ and $\mathsf{I}_{nneg(V)}$ is a stable model of $nneg(F)$ relative to $\sigma \cup \{np \mid p \in \sigma\}$.*

(b) An interpretation I of $\sigma \cup \{np \mid p \in \sigma\}$ is a stable model of $nneg(F)$ relative to $\sigma \cup \{np \mid p \in \sigma\}$ iff $I = I_{nneg(V)}$ for some fuzzy equilibrium model V of F.

Example 10. *For fuzzy formula $F = (0.2 \to_r p) \otimes_m (0.3 \to_r {\sim} p)$, formula $nneg(F)$ is*

$$(0.2 \to_r p) \otimes_m (0.3 \to_r np) \otimes_m \neg_s (p \otimes_l np).$$

One can check that the valuation V defined as $V(w,p) = [0.2, 0.7]$ is the only equilibrium model of F, and the interpretation $I_{nneg(V)} = \{(p, 0.2), (np, 0.3)\}$ is the only fuzzy stable model of $nneg(F)$.

This idea of eliminating strong negation in favor of new atoms was used in Examples 3 and 4.

The correspondence between fuzzy equilibrium models and fuzzy stable models indicates that the complexity analysis for fuzzy equilibrium logic applies to fuzzy stable models as well. In [25] it is shown that deciding whether a formula has a fuzzy equilibrium model is Σ_2^P-hard, which applies to the fuzzy stable model semantics as well. The same problem for a normal FASP programs, which can be identified with a special case of the fuzzy stable model semantics as shown in Section 6.1, is NP-hard. Complexity analyses for other special cases based on restrictions on fuzzy operators, or the presence of cycles in a program have been studied in [3].

7 Some Properties of Fuzzy Stable Models

In this section, we show that several well-known properties of the Boolean stable model semantics can be naturally extended to the fuzzy stable model semantics.

7.1 Theorem on Constraints

In answer set programming, constraints—rules with $\bot$ in the head—play an important role in view of the fact that adding a constraint eliminates the stable models that "violate" the constraint. The following theorem is the counterpart of Theorem 3 from [9] for fuzzy propositional formulas.

Theorem 8. *For any fuzzy formulas F and G, I is a stable model of $F \otimes \neg G$ (relative to $\mathbf{p}$) if and only if I is a stable model of F (relative to $\mathbf{p}$) and $I \models \neg G$.*

Example 11. *Consider $F = (\neg_s p \to_r q) \otimes_m (\neg_s q \to_r p) \otimes_m \neg_s p$. Formula F has only one stable model $I = \{(p, 0), (q, 1)\}$, which is the only stable model of $(\neg_s p \to_r q) \otimes_m (\neg_s q \to_r p)$ that satisfies $\neg_s p$.*

7.2 Theorem on Choice Formulas

In the Boolean stable model semantics, formulas of the form $p \vee \neg p$ are called *choice formulas*, and adding them to the program makes atoms p exempt from minimization. Choice formulas have been shown to be useful in constructing an ASP program in the "Generate-and-Test" style. This section shows their counterpart in the fuzzy stable model semantics.

For any finite set of fuzzy atoms $\mathbf{p} = \{p_1, \ldots, p_n\}$, the expression $\mathbf{p}^{ch}$ stands for the choice formula

$$(p_1 \oplus_l \neg_s p_1) \otimes \ldots \otimes (p_n \oplus_l \neg_s p_n),$$

where $\otimes$ is any fuzzy conjunction.

The following proposition tells us that choice formulas are tautological.

Proposition 4. *For any fuzzy interpretation I and any finite set $\mathbf{p}$ of fuzzy atoms, $I \models \mathbf{p}^{ch}$.*[9]

Theorem 9 is an extension of Theorem 2 from [9].

Theorem 9. *(a) For any real number $y \in [0,1]$, if I is a y-stable model of F relative to $\mathbf{p} \cup \mathbf{q}$, then I is a y-stable model of F relative to $\mathbf{p}$.*

(b) I is a 1-stable model of F relative to $\mathbf{p}$ iff I is a 1-stable model of $F \otimes \mathbf{q}^{ch}$ relative to $\mathbf{p} \cup \mathbf{q}$.

Theorem 9 (b) does not hold for an arbitrary threshold y (i.e., if "1−" is replaced with "$y-$"). For example, consider $F = \neg_s \neg_s q$ and $I = \{(q, 0.5)\}$. Clearly, I is a 0.5-model of F, and thus I is a 0.5-stable model of F relative to $\emptyset$. However, I is not a 0.5-stable model of $F \otimes_m \{q\}^{ch} = \neg_s \neg_s q \otimes_m (q \oplus_l \neg_s q)$ relative to $\emptyset \cup \{q\}$, as witnessed by $J = \{(q,0)\}$.

Since the 1-stable models of F relative to $\emptyset$ are the models of F, it follows from Theorem 9 (b) that the 1-*stable models* of $F \otimes \sigma^{ch}$ relative to the whole signature σ are exactly the 1-*models* of F.

Corollary 2. *Let F be a fuzzy formula of a finite signature σ. I is a model of F iff I is a stable model of $F \otimes \sigma^{ch}$ relative to σ.*

Example 12. *Consider the fuzzy formula $F = \neg_s q \rightarrow_r p$ in Example 1, which has only one stable model $\{(p,1),(q,0)\}$, although any interpretation $I = \{(p,x),(q,1-$*

---

[9] This proposition may not hold if $\oplus_l$ in the choice formula is replaced by an arbitrary fuzzy disjunction. For example, consider using $\oplus_m$ instead. Clearly, the interpretation $I = \{(p, 0.5)\} \not\models p \oplus_m \neg_s p$.

$x)\}$ is a model of F. In accordance with Corollary 2, we check that any I is a stable model of $G = F \otimes_m (p \oplus_l \neg_s p) \otimes_m (q \oplus_l \neg_s q)$. The reduct G^I is equivalent to

$$((1 - v_I(q)) \to_r p) \otimes_m (p \oplus_l (1 - v_I(p))) \otimes_m (q \oplus_l (1 - v_I(q))).$$

It is clear that any interpretation J that satisfies G^I should be such that $v_J(p) \geq v_I(p)$ and $v_J(q) \geq v_I(q)$, so there is no witness to dispute the stability of I.

8 Other Related Work

Several approaches to incorporating graded truth degrees into the answer set programming framework have been proposed. In this paper, we have formally compared our approach to [25] and [19]. Most works consider a special form $h \leftarrow B$ where h is an atom and B is some formula [4, 5, 23, 28]. Among them, [5, 23, 28] allow B to be any arbitrary formula corresponding to an increasing function whose arguments are the atoms appearing in the formula. [4] allows B to correspond to either an increasing function or a decreasing function. [20] considers the normal program syntax, i.e., each rule is of the form $l_0 \leftarrow l_1 \otimes \ldots \otimes l_m \otimes not\, l_{m+1} \otimes \ldots \otimes not\, l_n$, where each l_i is an atom or a strongly negated atom. In terms of semantics, most of the previous works rely on the notion of an immediate consequence operator and relate the fixpoint of this operator to the minimal model of a positive program. Similar to the approach [19] has adopted, the answer set of a positive program is defined as its minimal model, while the answer sets of a non-positive program are defined as minimal models of reducts. [27] presented a semantics based on the notion of an unfounded set. Disjunctive fuzzy answer set programs were also studied in [3].

It is worth noting that some of the related works have discussed so-called residuated programs [5, 20, 23, 28], where each rule $h \leftarrow B$ is assigned a weight θ, and a rule is satisfied by an interpretation I if $I(h \leftarrow B) \geq \theta$. According to [5], this class of programs is able to capture many other logic programming paradigms, such as possibilistic logic programming, hybrid probabilistic logic programming, and generalized annotated logic programming. Furthermore, as shown in [5], a weighted rule $(h \leftarrow B, \theta)$ can be simulated by $h \leftarrow B \otimes \theta$, where $(\otimes, \leftarrow)$ forms an adjoint pair. Notice that a similar method was used in Theorem 1 in relating y-stable models to 1-stable models.

It is well known in the Boolean stable model semantics that strong negation can be represented in terms of new atoms [9]. Our adaptation in the fuzzy stable model semantics is similar to the method from [20], in which the consistency of an interpretation is guaranteed by imposing the extra restriction $I(\sim p) \leq \sim I(p)$ for all atom p. Strong negation and consistency have also been studied in [21, 22].

In addition to fuzzy answer set programming, there are other approaches developed to handle many-valued logic. For example, [26] proposed a logic programming framework where each literal is annotated by a real-valued interval. Another example is possibilistic logic [6], where each propositional symbol is associated with two real values in the interval [0, 1] called the necessity degree and the possibility degree. Although these semantics handle fuzziness based on quite different ideas, it has been shown that these paradigms can be captured by fuzzy answer set programs [5].

While the development of FASP solvers has not yet reached the maturity level of ASP solvers, there is an increasing interest recently. [1] presented an FASP solver based on answer set approximation operators and a translation to bilevel linear programming presented in [3]. The implementation in [24] is based on a reduction of FASP to ASP. Independent from the work presented here, [2] presented another promising FASP solver, named FASP2SMT, that uses SMT solvers based on a translation from FASP into satisfiability modulo theories. A large fragment of the language proposed in this paper can be computed by this solver. The input language allows $\neg_s$, $\otimes_l$, $\oplus_l$, $\otimes_m$, $\oplus_m$ as fuzzy operators, and rules of the form

$$Head \leftarrow Body$$

where $Head$ is $p_1 \odot \cdots \odot p_n$ where p_i are atoms or numeric constants, and $\odot \in \{\otimes_l, \oplus_l, \otimes_m, \oplus_m\}$, and $Body$ is a nested formula formed from atoms and numeric constants using $\neg_s$, $\otimes_l$, $\oplus_l$, $\otimes_m$, $\oplus_m$.

Example 4 can be computed by FASP2SMT. Assume a conflict of degree 0.1 between Alice and Bob occurred at step 0, no conflict occurred at step 1, and a conflict of degree 0.5 between Alice and Bob occurred at step 2. Since the current version of `fast2smt` [10] does not yet support product conjunction $\otimes_p$, we use Łukasiewicz conjunction $\otimes_l$ in formula F_4. In the input language of FASP2SMT, ":-" denote $\rightarrow_r$, "," denotes $\otimes_m$, "*" denote $\otimes_l$, "+" denotes $\oplus_l$, and "not" denotes $\neg_s$. Numeric constants begin with "#" symbol, and variables are capitalized. The encoding in the input language of FASP2SMT is shown in Figure 2.

The command line to compute this program is simply:

```
python fasp2smt.py trust.fasp
```

Part of the output from FASP2SMT is shown below:

trust(alice,bob,0)	0.800000	(4.0/5.0)
trust(bob,carol,0)	0.700000	(7.0/10.0)
trust(alice,carol,0)	0.500000	(1.0/2.0)
trust(alice,bob,1)	0.700000	(7.0/10.0)

---

[10] Downloaded in March 2016

```
% domain
person(alice).      person(bob).       person(carol).
step(0).            step(1).           step(2).          step(3).
next(0,1).          next(1,2).         next(2,3).

% F1: trust is reflexive
trust(X,X,T) :- person(X), step(T).

% F2, F3: UEC
:- trust(X,Y,T) * distrust(X,Y,T).
:- not (trust(X,Y,T) + distrust(X,Y,T)), person(X), person(Y), step(T).

% F4, F5: transitivity of trust
% F4 modified: product t-norm is replaced by Lukasiwicz t-norm since
%   the solver does not support it.
trust(X,Z,0) :- trust(X,Y,0) * trust(Y,Z,0),
                person(X), person(Y), person(Z).
distrust(X,Y,0) :- not trust(X,Y,0), person(X), person(Y).

% F6, F7: inertia
trust(X,Y,T2) :- trust(X,Y,T1), not not trust(X,Y,T2), next(T1,T2).
distrust(X,Y,T2) :- distrust(X,Y,T1), not not distrust(X,Y,T2), next(T1,T2).

% F8: effect of conflict
distrust(X,Y,T2) :- conflict(X,Y,T1) + distrust(X,Y,T1), next(T1,T2).

% initial State
trust(alice,bob,0) :- #0.8.
trust(bob,carol,0) :- #0.7.

% action
conflict(alice,bob,0) :- #0.1.
conflict(alice,bob,2) :- #0.5.
```

Figure 2: Trust Example in the Input Language of FASP2SMT

trust(bob,carol,1)	0.700000	(7.0/10.0)
trust(alice,carol,1)	0.500000	(1.0/2.0)
trust(alice,bob,2)	0.700000	(7.0/10.0)
trust(bob,carol,2)	0.700000	(7.0/10.0)
trust(alice,carol,2)	0.500000	(1.0/2.0)
trust(alice,bob,3)	0.200000	(1.0/5.0)
trust(bob,carol,3)	0.700000	(7.0/10.0)
trust(alice,carol,3)	0.500000	(1.0/2.0)
conflict(alice,bob,0)	0.100000	(1.0/10.0)
conflict(alice,bob,1)	0.0	(0.0)
conflict(alice,bob,2)	0.500000	(1.0/2.0)

The number following each atom is the truth value of the atom in decimal notation, and the number in the parentheses is the truth value in fraction.

9 Conclusion

We introduced a stable model semantics for fuzzy propositional formulas, which generalizes both the Boolean stable model semantics and fuzzy propositional logic. The syntax is the same as the syntax of fuzzy propositional logic, but the semantics allows us to distinguish *stable models* from non-stable models. The formalism allows highly configurable default reasoning involving fuzzy truth values. The proposed semantics, when we restrict threshold to be 1 and assume all atoms to be subject to minimization, is essentially equivalent to fuzzy equilibrium logic, but is much simpler. To the best of our knowledge, our representation of the commonsense law of inertia involving fuzzy values is new. The representation uses nested fuzzy operators, which are not available in other fuzzy ASP semantics with restricted syntax.

We showed that several traditional results in answer set programming can be naturally extended to this formalism, and expect that more results can be carried over. Also, it would be possible to generalize the semantics to the first-order level, similar to the way the Boolean stable model semantics was generalized in [9].

References

[1] Mario Alviano and Rafael Penaloza. Fuzzy answer sets approximations. *Theory and Practice of Logic Programming*, 13(4-5):753–767, 2013.

[2] Mario Alviano and Rafael Penaloza. Fuzzy answer set computation via satisfiability modulo theories. *Theory and Practice of Logic Programming*, 15(4-5):588–603, 2015.

[3] Marjon Blondeel, Steven Schockaert, Dirk Vermeir, and Martine De Cock. Complexity of fuzzy answer set programming under Łukasiewicz semantics. *International Journal of Approximate Reasoning*, 55(9):1971–2003, 2014.

[4] Carlos Viegas Damásio and Luís Moniz Pereira. Antitonic logic programs. In *Proceedings of International Conference on Logic Programming and Nonmonotonic Reasoning (LPNMR)*, pages 379–392, 2001.

[5] Carlos Viegas Damásio and Luís Moniz Pereira. Monotonic and residuated logic programs. In Salem Benferhat and Philippe Besnard, editors, *ECSQARU*, volume 2143 of *Lecture Notes in Computer Science*, pages 748–759. Springer, 2001.

[6] Didier Dubois and Henri Prade. Possibilistic logic: a retrospective and prospective view. *Fuzzy Sets and Systems*, 144(1):3–23, 2004.

[7] Paolo Ferraris, Joohyung Lee, and Vladimir Lifschitz. A generalization of the Lin-Zhao theorem. *Annals of Mathematics and Artificial Intelligence*, 47:79–101, 2006.

[8] Paolo Ferraris, Joohyung Lee, Vladimir Lifschitz, and Ravi Palla. Symmetric splitting in the general theory of stable models. In *Proceedings of International Joint Conference on Artificial Intelligence (IJCAI)*, pages 797–803. AAAI Press, 2009.

[9] Paolo Ferraris, Joohyung Lee, and Vladimir Lifschitz. Stable models and circumscription. *Artificial Intelligence*, 175:236–263, 2011.

[10] Paolo Ferraris. Answer sets for propositional theories. In *Proceedings of International Conference on Logic Programming and Nonmonotonic Reasoning (LPNMR)*, pages 119–131, 2005.

[11] Peter Hajek. *Mathematics of Fuzzy Logic*. Kluwer, 1998.

[12] Jeroen Janssen, Dirk Vermeir, Steven Schockaert, and Martine De Cock. Reducing fuzzy answer set programming to model finding in fuzzy logics. *Theory and Practice of Logic Programming*, 12(06):811–842, 2012.

[13] Joohyung Lee and Yi Wang. Stable models of fuzzy propositional formulas. In *Working Notes of the 1st Workshop on Logics for Reasoning about Preferences, Uncertainty, and Vagueness (PRUV)*, 2014.

[14] Joohyung Lee and Yi Wang. Stable models of fuzzy propositional formulas. In *Proceedings of European Conference on Logics in Artificial Intelligence (JELIA)*, pages 326–339, 2014.

[15] Joohyung Lee, Vladimir Lifschitz, and Ravi Palla. A reductive semantics for counting and choice in answer set programming. In *Proceedings of the AAAI Conference on Artificial Intelligence (AAAI)*, pages 472–479, 2008.

[16] Vladimir Lifschitz and Hudson Turner. Representing transition systems by logic programs. In *Proceedings of International Conference on Logic Programming and Nonmonotonic Reasoning (LPNMR)*, pages 92–106, 1999.

[17] Vladimir Lifschitz. Answer set programming and plan generation. *Artificial Intelligence*, 138:39–54, 2002.

[18] Vladimir Lifschitz. What is answer set programming? In *Proceedings of the AAAI Conference on Artificial Intelligence*, pages 1594–1597. MIT Press, 2008.

[19] Thomas Lukasiewicz. Fuzzy description logic programs under the answer set semantics for the semantic web. In Thomas Eiter, Enrico Franconi, Ralph Hodgson, and Susie Stephens, editors, *RuleML*, pages 89–96. IEEE Computer Society, 2006.

[20] Nicolás Madrid and Manuel Ojeda-Aciego. Towards a fuzzy answer set semantics for residuated logic programs. In *Web Intelligence/IAT Workshops*, pages 260–264. IEEE, 2008.

[21] Nicolás Madrid and Manuel Ojeda-Aciego. On coherence and consistence in fuzzy answer set semantics for residuated logic programs. In Vito Di Gesù, Sankar K. Pal, and Alfredo Petrosino, editors, *WILF*, volume 5571 of *Lecture Notes in Computer Science*, pages 60–67. Springer, 2009.

[22] Nicolás Madrid and Manuel Ojeda-Aciego. Measuring inconsistency in fuzzy answer set semantics. *IEEE T. Fuzzy Systems*, 19(4):605–622, 2011.

[23] Jesús Medina, Manuel Ojeda-Aciego, and Peter Vojtás. Multi-adjoint logic programming with continuous semantics. In *Proceedings of International Conference on Logic Programming and Nonmonotonic Reasoning (LPNMR)*, pages 351–364, 2001.

[24] Mushthofa Mushthofa, Steven Schockaert, and Martine De Cock. A finite-valued solver for disjunctive fuzzy answer set programs. In *21st European conference on Artificial Intelligence (ECAI 2014)*, volume 263, pages 645–650. IOS Press, 2014.

[25] Steven Schockaert, Jeroen Janssen, and Dirk Vermeir. Fuzzy equilibrium logic: Declarative problem solving in continuous domains. *ACM Transactions on Computational Logic (TOCL)*, 13(4):33, 2012.

[26] Umberto Straccia. Annotated answer set programming. In *Proceedings of the 11th International Conference on Information Processing and Management of Uncertainty in Knowledge-Based Systems*, pages 1212–1219, 2006.

[27] Davy Van Nieuwenborgh, Martine De Cock, and Dirk Vermeir. An introduction to fuzzy answer set programming. *Annals of Mathematics and Artificial Intelligence*, 50(3-4):363–388, 2007.

[28] Peter Vojtás. Fuzzy logic programming. *Fuzzy Sets and Systems*, 124(3):361–370, 2001.

A Proofs

A.1 Proof of Proposition 1

Lemma 2. *For any fuzzy conjunction $\otimes$, we have $\otimes(x,y) \leq x$ and $\otimes(x,y) \leq y$.*

Proof. By the conditions imposed on fuzzy conjunctions, we have $\otimes(x,y) \leq \otimes(x,1) = x$ and $\otimes(x,y) \leq \otimes(1,y) = y$. ∎

Proposition 1. *For any interpretations I and J such that $J \leq^{\mathbf{p}} I$, it holds that*

$$v_J(F^I) \leq v_I(F).$$

Proof. By induction on F.

- F is an atom p. $v_J(F^I) = J(p)$ and $v_I(F) = I(p)$. Clear from the assumption $J \leq^p I$.

- F is a numeric constant c. Clearly, $v_J(F^I) = c = v_I(F)$.

- F is $\neg G$. $v_J(F^I) = v_J(v_I(\neg G)) = v_I(\neg G) = v_I(F)$.

- F is $G \odot H$, where $\odot$ is $\otimes$ or $\oplus$. $v_J(F^I) = v_J(G^I) \odot v_J(H^I)$. By I.H. we have $v_J(G^I) \leq v_I(G)$ and $v_J(H^I) \leq v_I(H)$. Since $\otimes$ and $\oplus$ are both increasing, we have $v_J(F^I) = v_J(G^I) \odot v_J(H^I) \leq v_I(G) \odot v_I(H) = v_I(F)$.

- F is $G \to H$. $v_J(F^I) = (v_J(G^I) \to v_J(H^I)) \otimes_m v_I(F)$. By Lemma 2, $v_J(F^I) \leq v_I(F)$.

∎

A.2 Proof of Proposition 2

Lemma 3. *For any (fuzzy) formula F and (fuzzy) interpretation I, we have $v_I(F) = v_I(F^I)$.*

Proof. By induction on F.

- F is an atom p or a numeric constant. Clear from the fact $F^I = F$.

- F is $\neg G$. Then we have $v_I(F) = v_I(\neg G) = v_I(v_I(\neg G)) = v_I(F^I)$.

- F is $G \odot H$, where $\odot$ is $\otimes$, $\oplus$, or $\to$. Then $F^I = (G^I \odot H^I) \otimes_m v_I(F \odot G)$. By I.H., we have $v_I(G) = v_I(G^I)$ and $v_I(H) = v_I(H^I)$. So we have

$$v_I(F^I) = \min\left\{v_I(G^I \odot H^I),\ v_I(G \odot H)\right\}$$
$$= \min\left\{\odot(v_I(G^I), v_I(H^I)),\ \odot(v_I(G), v_I(H))\right\}$$
$$= \min\{\odot(v_I(G), v_I(H)),\ \odot(v_I(G), v_I(H))\}$$
$$= \odot(v_I(G), v_I(H))$$
$$= v_I(F).$$

∎

Proposition 2 is an immediate corollary to Lemma 3.

Proposition 2. *A (fuzzy) interpretation I satisfies a (fuzzy) formula F if and only if I satisfies F^I.*

A.3 Proof of Theorem 1

Lemma 4. *For any fuzzy formula F, any interpretation I, and any implication $\rightarrow$ that satisfies $\rightarrow(x,y) = 1$ iff $y \geq x$, we have that I is a y-model of F iff I is a 1-model of $y \rightarrow F$.*

Proof. By definition, $I \models_y F$ means that $v_I(F) \geq y$. Since $\rightarrow(x,y) = 1$ iff $y \geq x$, $v_I(F) \geq y$ iff $\rightarrow(y, v_I(F)) = 1$ iff $v_I(y \rightarrow F) = 1$ iff $I \models_1 (y \rightarrow F)$. ∎

Theorem 1. *For any fuzzy formula F, an interpretation I is a y-stable model of F relative to $\mathbf{p}$ iff I is a 1-stable model of $y \rightarrow F$ relative to $\mathbf{p}$ as long as the implication $\rightarrow$ satisfies the condition $\rightarrow(x,y) = 1$ iff $y \geq x$.*

Proof.

I is a y-stable model of F relative to $\mathbf{p}$

iff

$I \models_y F$ and there is no $J <^{\mathbf{p}} I$ such that $J \models_y F^I$

iff (by Lemma 4)

$I \models_1 y \rightarrow F$ and there is no $J <^{\mathbf{p}} I$ such that $J \models_1 (y \rightarrow F^I)$

iff (since $v_I(y \rightarrow F) = 1$)

$I \models_1 y \rightarrow F$ and there is no $J <^{\mathbf{p}} I$ such that $J \models_1 (y \rightarrow F^I) \otimes_m v_I(y \rightarrow F)$

iff (since $(y \rightarrow F^I) \otimes_m v_I(y \rightarrow F) = (y \rightarrow F)^I$)

I is a 1-stable model of $y \rightarrow F$ relative to $\mathbf{p}$.

∎

A.4 Proof of Theorem 2

Lemma 5. *For any formula F and any interpretations I and J such that $J \leq^{\mathbf{p}} I$, $v_{I \cup J_{\mathbf{q}}^{\mathbf{p}}}(F^*(\mathbf{q})) = v_J(F^I)$.*

Proof. By induction on F.

- F is a numeric constant c, or an atom not in $\mathbf{p}$. Immediate from the fact that J and I agree on $F^*(\mathbf{q}) = F = F^I$.

- F is an atom $p_i \in \mathbf{p}$. $F^*(\mathbf{q}) = q_i$ and $F^I = p_i$. Clear from the fact that $v_{I \cup J_\mathbf{q}^\mathbf{p}}(q_i) = v_J(p_i)$.

- F is $\neg G$. $v_{I \cup J_\mathbf{q}^\mathbf{p}}((\neg G)^*(\mathbf{q})) = v_I(\neg G) = v_J(v_I(\neg G)) = v_J(F^I)$.

- F is $G \odot H$, where $\odot$ is $\otimes$ or $\oplus$. Immediate by I.H. on G and H.

- F is $G \to H$. By I.H., $v_{I \cup J_\mathbf{q}^\mathbf{p}}(G^*(\mathbf{q})) = v_J(G^I)$ and $v_{I \cup J_\mathbf{q}^\mathbf{p}}(H^*(\mathbf{q})) = v_J(H^I)$. $F^*(\mathbf{q})$ is $(G^*(\mathbf{q}) \to H^*(\mathbf{q})) \otimes_m (G \to H)$, and F^I is $(G^I \to H^I) \otimes_m v_I(G \to H)$. Then the claim is immediate from I.H.

∎

Theorem 2. *A fuzzy interpretation I is a fuzzy stable model of F relative to $\mathbf{p}$ iff*

- *$I \models F$, and*

- *there is no fuzzy interpretation J such that $J <^\mathbf{p} I$ and $I \cup J_\mathbf{q}^\mathbf{p} \models F^*(\mathbf{q})$.*

Proof.

I is a fuzzy stable model of F relative to $\mathbf{p}$

iff

$I \models F$ and there is no $J <^\mathbf{p} I$ such that $J \models F^I$

iff (by Lemma 5)

$I \models F$ and there is no $J <^\mathbf{p} I$ such that $I \cup J_\mathbf{q}^\mathbf{p} \models F^*(\mathbf{q})$.

∎

A.5 Proofs of Theorems 3 and 4

The following lemma immediately follows from Lemma 2.

Lemma 6. *For any fuzzy conjunction $\otimes$, $\otimes(x,y) = 1$ if and only if $x = 1$ and $y = 1$.*

Define the mapping $defuz(I)$ that maps a fuzzy interpretation $I = \{(p_1, x_1), \ldots, (p_n, x_n)\}$ to a classical interpretation such that $defuz(I) = \{p_i \mid (p_i, 1) \in I\}$.

Lemma 7. *for any fuzzy interpretation I and any classical propositional formula F,*

(i) *if $v_I(F^{fuzzy}) = 1$, then $defuz(I) \models F$, and*

(ii) if $v_I(F^{fuzzy}) = 0$, then $defuz(I) \not\models F$.

Proof. We prove by induction on F.

- F is an atom p. (i) Suppose $I \models_1 F^{fuzzy}$. Then $v_I(p) = 1$, and thus $p \in defuz(I)$. So $defuz(I) \models F$. (ii) Suppose $v_I(F^{fuzzy}) = 0$. Then $v_I(p) = 0$, and thus $p \notin defuz(I)$. So $defuz(I) \not\models F$.

- F is $\bot$. (i) There is no interpretation that satisfies $F^{fuzzy} = 0$ to the degree 1. So the claim is trivially true. (ii) Since no interpretation satisfies F, $defuz(I) \not\models F$.

- F is $\top$. (i) All interpretations satisfy F. So $defuz(I) \models F$. (ii) There is no interpretation I such that $v_I(F^{fuzzy}) = 0$. So (ii) is trivially true.

- F is $\neg G$. Then F^{fuzzy} is $\neg_s G^{fuzzy}$. (i) Suppose $I \models_1 F^{fuzzy}$. We have $v_I(F^{fuzzy}) = 1 - v_I(G^{fuzzy}) = 1$, so $v_I(G^{fuzzy}) = 0$. By I.H., $defuz(I) \not\models G$, and thus $defuz(I) \models F$. (ii) Suppose $v_I(F^{fuzzy}) = 0$, Then $1 - v_I(G^{fuzzy}) = 0$, $v_I(G^{fuzzy}) = 1$, i.e., $I \models_1 v_I(G^{fuzzy})$. By I.H., $defuz(I) \models G$, and thus $defuz(I) \not\models F$.

- F is $G \land H$. Then F^{fuzzy} is $G^{fuzzy} \otimes_m H^{fuzzy}$. (i) Suppose $I \models_1 F^{fuzzy}$. By Lemma 6, $v_I(G^{fuzzy}) = v_I(H^{fuzzy}) = 1$, i.e., $I \models_1 G^{fuzzy}$ and $I \models_1 H^{fuzzy}$. By I.H., $defuz(I) \models G$ and $defuz(I) \models H$. It follows that $defuz(I) \models G \land H = F$. (ii) Suppose $v_I(F^{fuzzy}) = min(v_I(G^{fuzzy}), v_I(H^{fuzzy})) = 0$. Then $v_I(G^{fuzzy}) = 0$ or $v_I(H^{fuzzy}) = 0$. By I.H., $defuz(I) \not\models G$ or $defuz(I) \not\models H$. It follows that $defuz(I) \not\models G \land H = F$.

- F is $G \lor H$. Then F^{fuzzy} is $G^{fuzzy} \oplus_m H^{fuzzy}$. (i) Suppose $I \models_1 F^{fuzzy}$, and as the disjunction is defined as $\oplus_m(x,y) = max(x,y)$, $v_I(G^{fuzzy}) = 1$ or $v_I(H^{fuzzy}) = 1$, i.e., $I \models_1 G^{fuzzy}$ or $I \models_1 H^{fuzzy}$. By I.H., $defuz(I) \models G$ or $defuz(I) \models H$. It follows that $defuz(I) \models G \lor H = F$. (ii) Suppose $v_I(F^{fuzzy}) = max(v_I(G^{fuzzy}), v_I(H^{fuzzy})) = 0$. Then $v_I(G^{fuzzy}) = 0$ and $v_I(H^{fuzzy}) = 0$. By I.H., $defuz(I) \not\models G$ and $defuz(I) \not\models H$. It follows that $defuz(I) \not\models G \lor H = F$.

- F is $G \to H$. Then F^{fuzzy} is $G^{fuzzy} \to_s H^{fuzzy}$. (i) Suppose $I \models_1 F^{fuzzy}$. We have $v_I(F^{fuzzy}) = max(1 - v_I(G^{fuzzy}), v_I(H^{fuzzy})) = 1$, so that $v_I(G^{fuzzy}) = 0$ or $v_I(H^{fuzzy}) = 1$.[11] By I.H., $defuz(I) \not\models G$ or $defuz(I) \models H$. It follows

---

[11]This does not hold for an arbitrary choice of implication. For example, consider $\to_l (x,y) = min(1-x+y, 1)$, then from $I \models_1 G \to_l H$, we can only conclude $v_I(H) \geq v_I(G)$. Furthermore, under this choice of implication, the modified statement of the lemma does not hold. A counterexample is $F = \neg_s p \to_l q$, $I = \{(p, 0.5), (q, 0.6)\}$. Clearly, $I \models_1 F^{fuzzy}$ but $defuz(I) = \emptyset \not\models F$.

that $defuz(I) \models G \rightarrow H = F$.

(ii) Suppose $v_I(F^{fuzzy}) = max\{1 - v_I(G^{fuzzy}), v_I(H^{fuzzy})\} = 0$. Then $v_I(G^{fuzzy}) = 1$ and $v_I(H^{fuzzy}) = 0$. By I.H., $defuz(I) \models G$ and $defuz(I) \not\models H$. Therefore $defuz(I) \not\models G \rightarrow H$. ∎

Lemma 8. *For any classical interpretation I and any classical propositional formula F, $I \models F$ if and only if $I^{fuzzy} \models F^{fuzzy}$.*

Proof. By induction on F. ∎

Theorem 3. *For any classical propositional formula F and any classical propositional interpretation I, I is a stable model of F relative to $\mathbf{p}$ iff I^{fuzzy} is a stable model of F^{fuzzy} relative to $\mathbf{p}$.*

Proof. ($\Rightarrow$) Suppose I is a stable model of F relative to $\mathbf{p}$. From the fact that $I \models F$, by Lemma 8, $I^{fuzzy} \models F^{fuzzy}$.

Next we show that there is no fuzzy interpretation $J <^{\mathbf{p}} I^{fuzzy}$ such that $J \models (F^{fuzzy})^I$. Suppose, for the sake of contradiction, that there exists such J. Since $J \models (F^{fuzzy})^I$, or equivalently, $J \models (F^I)^{fuzzy}$, by Lemma 7, we get $defuz(J) \models F^I$.

Since $J <^{\mathbf{p}} I^{fuzzy}$, J and I^{fuzzy} agree on all atoms not in $\mathbf{p}$. Since I^{fuzzy} assigns either 0 or 1 to each atom, it follows that J, as well as $defuz(J)$, assigns the same truth values as I^{fuzzy} to atoms not in $\mathbf{p}$. The construction of $defuz(J)$ guarantees that $defuz(J)^{fuzzy} \leq^{\mathbf{p}} J <^{\mathbf{p}} I^{fuzzy}$. Since both $defuz(J)^{fuzzy}$ and I^{fuzzy} assigns either 0 or 1 to each atom, there is at least one atom $p \in \mathbf{p}$ such that $v_{defuz(J)^{fuzzy}}(p) = 0$ and $v_{I^{fuzzy}}(p) = 1$, and consequently $defuz(J)(p) = $ FALSE and $I(p) = $ TRUE. So $defuz(J) <^{\mathbf{p}} I$, and together with the fact that $defuz(J) \models F^I$, it follows that I is not a stable model of F, which is a contradiction. Thus, there is no such J, from which we conclude that I^{fuzzy} is a stable model of F^{fuzzy} relative to $\mathbf{p}$.

($\Leftarrow$) Suppose I^{fuzzy} is a stable model of a fuzzy formula F^{fuzzy} relative to $\mathbf{p}$. Then $I^{fuzzy} \models F^{fuzzy}$. By Lemma 8, $I \models F$.

Next we show there is no $J <^{\mathbf{p}} I$ such that $J \models F^I$. Suppose, for the sake of contradiction, that there exists such J. Then by Lemma 8, $J^{fuzzy} \models (F^I)^{fuzzy}$, and obviously $J^{fuzzy} <^{\mathbf{p}} I^{fuzzy}$. It follows that I^{fuzzy} is not a stable model of F^{fuzzy} relative to $\mathbf{p}$, which is a contradiction. So there is no such J, from which we conclude that I is a stable model of F relative to $\mathbf{p}$. ∎

The proof of Theorem 4 is the same as the right-to-left direction of the proof of Theorem 3. Notice that the left-to-right direction of the proof of Theorem 3 relies on Lemma 7, which assumes the particular selection of fuzzy operators, so this direction does not apply to the setting of Theorem 4.

A.6 Proof of Theorem 5

Theorem 5. *For any normal FASP program $\Pi = \{r_1, \ldots, r_n\}$, let F be the fuzzy formula $r_1 \otimes \ldots \otimes r_n$, where $\otimes$ is any fuzzy conjunction. An interpretation I is a fuzzy answer set of Π in the sense of [19] if and only if I is a stable model of F.*

Proof. By Π^I we denote the reduct of Π relative to I as defined in [19], which is also reviewed in Section 6.1.

($\Rightarrow$) Suppose I is a fuzzy answer set of Π. By definition, $I \models \Pi$, and thus $v_I(r_i) = 1$ for all $r_i \in \Pi$. So
$$v_I(F) = v_I(r_1 \otimes \ldots \otimes r_n) = 1,$$
i.e., $I \models F$.

Next we show that there is no $J <^\sigma I$ such that $J \models F^I$, where σ is the underlying signature. For each $r_i = a \leftarrow_r b_1 \otimes \ldots \otimes b_m \otimes \neg b_{m+1} \otimes \ldots \otimes \neg b_n$,
$$r_i^I = v_I(r_i) \otimes_m (a \leftarrow_r b_1 \otimes \ldots \otimes b_m \otimes v_I(\neg_s b_{m+1}) \otimes \ldots \otimes v_I(\neg_s b_n)).$$

Suppose, for the sake of contradiction, that there exists an interpretation $J <^\sigma I$ such that $J \models F^I$. Then, for all $r_i \in \Pi$, $J \models r_i^I$, i.e.,
$$J \models v_I(r_i) \otimes_m (a \leftarrow_r b_1 \otimes \ldots \otimes b_m \otimes v_I(\neg_s b_{m+1}) \otimes \ldots \otimes v_I(\neg_s b_n)).$$

It follows that
$$J \models a \leftarrow_r b_1 \otimes \ldots \otimes b_m \otimes v_I(\neg_s b_{m+1}) \otimes \ldots \otimes v_I(\neg_s b_n).$$

So $J \models \Pi^I$. Together with the fact $J <^\sigma I$, this contradicts that I is a fuzzy answer set of Π. So there is no such J, from which we conclude that I is a stable model of F.

($\Leftarrow$) Suppose I is a stable model of F. From the fact that $I \models F$, it follows that $v_I(r_i) = 1$ for all $r_i \in \Pi$. Thus $I \models \Pi$.

Next we show that there is no $J <^\sigma I$ such that $J \models \Pi^I$, where σ is the underlying signature. The reduct Π^I contains the following rule for each original rule $r_i \in \Pi$:
$$a \leftarrow b_1 \otimes \ldots \otimes b_m \otimes v_I(\neg b_{m+1}) \otimes \ldots \otimes v_I(\neg b_n).$$

Suppose, for the sake of contradiction, that there exists $J <^\sigma I$ such that $J \models \Pi^I$. Then for each rule $r_i \in \Pi$,

$$J \models a \leftarrow_r b_1 \otimes \ldots \otimes b_m \otimes v_I(\neg_s b_{m+1}) \otimes \ldots \otimes v_I(\neg_s b_n).$$

As $I \models \Pi$, for all $r_i \in \Pi$, $v_I(r_i) = 1$, so

$$J \models v_I(r_i) \otimes_m (a \leftarrow_r b_1 \otimes \ldots \otimes b_m \otimes \neg_s v_I(b_{m+1}) \otimes \ldots \otimes v_I(\neg_s b_n))$$

or equivalently, $J \models r_i^I$ for each $r_i \in \Pi$. Therefore, $J \models (r_1 \otimes \ldots \otimes r_n)^I$, i.e., $J \models F^I$. Together with the fact $J <^\sigma I$, this contradicts that I is a stable model of F. So there is no such J, from which we conclude that I is a fuzzy answer set of Π. ∎

A.7 Proof of Theorem 6

Let σ be a signature, and let F be a fuzzy formula of signature σ that does not contain strong negation.

For any two interpretations I and J such that $J \leq^\sigma I$, we define the fuzzy N5 valuation $\mathsf{V}_{J,I}$ as follows. For every atom $p \in \sigma$,

- $\mathsf{V}_{J,I}(h, p) = [v_J(p), 1]$, and
- $\mathsf{V}_{J,I}(t, p) = [v_I(p), 1]$.

It can be seen that $\mathsf{V}_{J,I}$ is always a valid valuation as long as $J \leq^\sigma I$. Clearly, $\mathsf{I}_{\mathsf{V}_{I,I}} = I$ for any interpretation I.

Lemma 9. *For any interpretations I and J such that $J \leq^\sigma I$, it holds that $v_I(F) = \mathsf{V}^-_{J,I}(t, F)$.*

Proof. By induction on F.

- F is an atom p. Then $v_I(F) = v_I(p) = \mathsf{V}^-_{J,I}(t, p) = \mathsf{V}^-_{J,I}(t, F)$.
- F is a numeric constant c. Then $v_I(F) = c = \mathsf{V}^-_{J,I}(t, c) = \mathsf{V}^-_{J,I}(t, F)$.
- F is $\neg G$. Then $v_I(F) = \neg(v_I(G))$. By I.H., $v_I(F) = \neg(\mathsf{V}^-_{J,I}(t, G)) = \mathsf{V}^-_{J,I}(t, F)$.
- F is $G \odot H$, where $\odot$ is $\otimes$, $\oplus$ or $\rightarrow$. Then $v_I(F) = v_I(G) \odot v_I(H)$. By I.H., $v_I(F) = \mathsf{V}^-_{J,I}(t, G) \odot \mathsf{V}^-_{J,I}(t, H) = \mathsf{V}^-_{J,I}(t, F)$.

∎

The following is a corollary to Lemma 9.

Corollary 3. $I \models F$ if and only if $\mathsf{V}_{I,I}$ is a model of F.

Proof. By Lemma 9, $v_I(F) = \mathsf{V}^-_{I,I}(t, F)$. Furthermore, $\mathsf{V}^-_{I,I}(h, F) = \mathsf{V}^-_{I,I}(t, F)$ can be proven by induction. ∎

Lemma 10. For any interpretations I and J such that $J \leq^\sigma I$, it holds that $v_J(F^I) = \mathsf{V}^-_{J,I}(h, F)$.

Proof. By induction on F.

- F is an atom p or a numeric constant. Then $v_J(F^I) = v_J(F) = \mathsf{V}^-_{J,I}(h, F)$.

- F is $\neg G$. Then $v_J(F^I) = \neg v_I(G)$, and by Lemma 9, $\neg v_I(G) = \neg \mathsf{V}^-_{J,I}(t, G) = \mathsf{V}^-_{J,I}(h, F)$.

- F is $G \odot H$, where $\odot$ is $\otimes$ or $\oplus$. Then

$$\begin{aligned} v_J(F^I) &= v_J(G^I \odot H^I) = v_J(G^I) \odot v_J(H^I) \\ &= \text{(by I.H.)} \\ & \quad \mathsf{V}^-_{J,I}(h, G) \odot \mathsf{V}^-_{J,I}(h, H) \\ &= \mathsf{V}^-_{J,I}(h, G \odot H) = \mathsf{V}^-_{J,I}(h, F). \end{aligned}$$

- F is $G \to H$. Then

$$\begin{aligned} v_J(F^I) &= v_J((G^I \to H^I) \otimes_m v_I(G \to H)) \\ &= min(v_J(G^I \to H^I), v_I(G \to H)) \\ &= \text{(by I.H. and Lemma 9)} \\ & \quad min(\mathsf{V}^-_{J,I}(h, G) \to \mathsf{V}^-_{J,I}(h, H), \mathsf{V}^-_{J,I}(t, G) \to \mathsf{V}^-_{J,I}(t, H)) \\ &= \mathsf{V}^-_{J,I}(h, G \to H) = \mathsf{V}^-_{J,I}(h, F). \end{aligned}$$

∎

The following is a corollary to Lemma 10.

Corollary 4. For any interpretations I and J such that $J \leq^\sigma I$, $J \models F^I$ if and only if $\mathsf{V}_{J,I}$ is a model of F.

Lemma 11. For two interpretations I and J, it holds that $\mathsf{V}_{J,I} \prec \mathsf{V}_{I,I}$ if and only if $J < I$.

Proof. ($\Rightarrow$) Suppose $V_{J,I} \prec V_{I,I}$. For every atom a, $V_{I,I}(h,a) \subseteq V_{J,I}(h,a)$, which means $V^-_{J,I}(h,a) \le V^-_{I,I}(h,a)$. So $J \le I$. Furthermore, there is at least one atom p satisfying $V_{I,I}(h,p) \subset V_{J,I}(h,p)$. By the definition of $V_{I,I}$ and $V_{J,I}$, $V^+_{I,I}(h,a) = V^+_{J,I}(h,a) = 1$ for all a. Consequently, $V^-_{J,I}(h,p) < V^-_{I,I}(h,p)$, which means $v_J(p) < v_I(p)$. So $J < I$.

($\Leftarrow$) Suppose $J < I$. Then for every atom a, $v_J(a) \le v_I(a)$. It follows that $V^-_{J,I}(h,a) \le V^-_{I,I}(h,a)$ for all a, and as $V^+_{I,I}(h,a) = V^+_{J,I}(h,a) = 1$ for all a, we conclude $V_{I,I}(h,a) \subseteq V_{J,I}(h,a)$. Clearly, $V_{I,I}(t,a) = V_{J,I}(t,a)$ by definition. Furthermore there is at least one atom p such that $v_J(p) < v_I(p)$, i.e., $V^-_{J,I}(h,a) < V^-_{I,I}(h,a)$. So $V_{J,I} \prec V_{I,I}$. ∎

Lemma 12. *An interpretation I is a stable model of F if and only if $V_{I,I}$ is a fuzzy equilibrium model of F.*

Proof. ($\Rightarrow$) Suppose I is a stable model of F. As I is a model of F, by Corollary 3, $V_{I,I}$ is a model of F. Next we show that there is no model V' of F such that $V' \prec V_{I,I}$. Suppose, for the sake of contradiction, that there exists such V'. Define an interpretation J as $v_J(a) = V'^-(h,a)$ for all atoms a. Obviously $V' = V_{J,I}$. As $V' = V_{J,I}$ is a model of F, by Corollary 4, $J \models F^I$. Furthermore, as $V' = V_{J,I} \prec V_{I,I}$, by Lemma 11, $J < I$. Consequently, I is not a 1-stable model of F, which is a contradiction. So there does not exist such V', from which we conclude that $V_{I,I}$ is an h-minimal model of F. Clearly, $V_{I,I}(h,a) = V_{I,I}(t,a)$ for all atoms a. So $V_{I,I}$ is an equilibrium model of F.

($\Leftarrow$) Suppose $V_{I,I}$ is an equilibrium model of F. As $V_{I,I}$ is a model of F, by Corollary 3, $I \models F$. Next we show that there is no $J <^\sigma I$ such that $J \models F^I$. Suppose, for the sake of contradiction, that there exists such J. Then by Corollary 4, the valuation $V_{J,I}$ is a model of F. Furthermore, by Lemma 11, $V_{J,I} \prec V_{I,I}$. Consequently, $V_{I,I}$ is not an h-minimal model of F, which contradicts that $V_{I,I}$ is an equilibrium model of F. So there does not exist such J, from which we conclude that I is a stable model of F. ∎

Lemma 13. *For any two valuations V and V' such that $V^-(w,a) = V'^-(w,a)$ ($w \in \{h,t\}$) for all atoms a, and any formula F containing no strong negation, V is a model of F iff V' is a model of F.*

Proof. We show by induction that $V^-(w,F) = V'^-(w,F)$, where $w \in \{h,t\}$.

- F is an atom p. $V^-(w,F) = V^-(w,p) = V'^-(w,p) = V^-(w,F)$.
- F is a numeric constant c. Clearly, $V^-(w,F) = c = V'^-(w,F)$.

1962

- F is $\neg G$. By I.H, $V^-(w,G) = V'^-(w,G)$.

$$V^-(w,F) = 1 - V^-(t,G) = 1 - V'^-(t,G) = V'^-(w,F).$$

- F is $G \odot H$ where $\odot$ is $\otimes$ or $\oplus$. By I.H, $V^-(w,G) = V'^-(w,G)$ and $V^-(w,H) = V'^-(w,H)$. So

$$\begin{aligned} V^-(w,F) &= V^-(w,G) \odot V^-(w,H) \\ &= V'^-(w,G) \odot V'^-(w,H) \\ &= V'^-(w,F). \end{aligned}$$

- F is $G \to H$. By I.H, $V^-(w,G) = V'^-(w,G)$ and $V^-(w,H) = V'^-(w,H)$. So

$$\begin{aligned} V^-(h,F) &= min((V^-(h,G) \to V^-(h,H)), (V^-(t,G) \to V^-(t,H))) \\ &= min((V'^-(h,G) \to V'^-(h,H)), (V'^-(t,G) \to V'^-(t,H))) \\ &= V'^-(h,F). \end{aligned}$$

And

$$\begin{aligned} V^-(t,F) &= (V^-(t,G) \to V^-(t,H)) \\ &= (V'^-(t,G) \to V'^-(t,H)) \\ &= V'^-(t,F). \end{aligned}$$

So $V^-(h,F) = V'^-(h,F)$ and thus $V^-(h,F) = 1$ if and only if $V'^-(h,F) = 1$, i.e, V is a model of F if and only if V' is a model of F. ∎

Lemma 1. *Given a formula F containing no strong negation, any equilibrium model V of F satisfies $V^+(h,a) = V^+(t,a) = 1$ for all atoms a.*

Proof. Assume that V is an equilibrium model of F. It follows that $V^+(h,a) = V^+(t,a)$. Furthermore, for the sake of contradiction, assume that $V^+(h,a) = V^+(t,a) = v < 1$. Define V'' as $V''^-(w,a) = V^-(w,a)$, $V''^+(t,a) = V^+(t,a)$ and $V''^+(h,a) = v'$ where $v' \in (v,1]$. Clearly, $V'' \prec V$ and by Lemma 13, V'' is a model of F. So V is not an h-minimal model of F, which contradicts the assumption that V is an equilibrium model of F. Therefore, there does not exist such V. ∎

Theorem 6. *Let F be a fuzzy propositional formula of σ that contains no strong negation.*

(a) *A valuation V of σ is a fuzzy equilibrium model of F iff $V^-(h,p) = V^-(t,p)$, $V^+(h,p) = V^+(t,p) = 1$ for all atoms p in σ and I_V is a stable model of F relative to σ.*

(b) An interpretation I of σ is a stable model of F relative to σ iff $I = \mathsf{I}_V$ for some fuzzy equilibrium model V of F.

Proof. (a) ($\Rightarrow$) Suppose V is an equilibrium model of F. By Lemma 1, $V^+(h,a) = V^+(t,a) = 1$ for all atoms a. And by the definition of fuzzy equilibrium model, $V^-(h,a) = V^-(t,a)$. It can be seen that $V_{\mathsf{I}_V,\mathsf{I}_V} = V$. Since $V_{\mathsf{I}_V,\mathsf{I}_V}$ is an equilibrium model of F, by Lemma 12, I_V is a stable model of F relative to σ.

($\Leftarrow$) Suppose $V^+(h,a) = V^+(t,a) = 1$, $V^-(h,a) = V^-(t,a)$ for all atoms a, and I_V is a stable model of F relative to σ. By Lemma 12, $V_{\mathsf{I}_V,\mathsf{I}_V}$ is an equilibrium model of F. Since $V^+(h,a) = V^+(t,a) = 1$, $V^-(h,a) = V^-(t,a)$ for all atoms a, it holds that $V_{\mathsf{I}_V,\mathsf{I}_V} = V$. So V is an equilibrium model of F.

(b) ($\Rightarrow$) Suppose I is a stable model of F relative to σ. By Lemma 12, $V_{I,I}$ is an equilibrium model of F. It can be seen that $I = \mathsf{I}_{V_{I,I}}$.

($\Leftarrow$) Take any fuzzy equilibrium model V of F and let $I = \mathsf{I}_V$. By the definition of fuzzy equilibrium model and Lemma 1, $V^-(h,a) = V^-(t,a)$ and $V^+(h,a) = V^+(t,a) = 1$ for all atoms a, which means $V_{\mathsf{I}_V,\mathsf{I}_V} = V$. So $V_{\mathsf{I}_V,\mathsf{I}_V}$ is an equilibrium model of F. By Lemma 12, I_V is a stable model, so I is a stable model. ∎

A.8 Proof of Theorem 7

Lemma 14. *For any fuzzy formula F of signature σ that may contain strong negation and for any valuation V, it holds that $V^-(w,F) = nneg(V)^-(w, nneg(F))$.*

Proof. First we show by induction that $V^-(w,F) = nneg(V)^-(w, F')$, where F' is defined as in Section 6.2.3.

- F is an atom p in σ. Clear.

- F is $\sim p$, where p is an atom in σ. Then F' is np.

$$V^-(w,F) = V^-(w, \sim p) = 1 - V^+(w,p)$$
$$= nneg(V)^-(w, np) = nneg(V)^-(w, F').$$

- F is a numeric constant c. Clear.

- F is $\neg G$. By I.H., $V^-(w, G) = nneg(V)^-(w, G')$.

$$\begin{aligned}
V^-(w, F) &= V^-(w, \neg G) \\
&= 1 - V^-(t, G) \\
&= 1 - nneg(V)^-(t, G') \\
&= nneg(V)^-(w, \neg G') \\
&= nneg(V)^-(w, F').
\end{aligned}$$

- F is $G \odot H$, where $\odot$ is $\otimes$ or $\oplus$. By I.H., $V^-(w, G) = nneg(V)^-(w, G')$ and $V^-(w, H) = nneg(V)^-(w, H')$.

$$\begin{aligned}
V^-(w, F) &= V^-(w, G \odot H) \\
&= \odot(V^-(w, G), V^-(w, H)) \\
&= \odot(nneg(V)^-(w, G'), nneg(V)^-(w, H')) \\
&= nneg(V)^-(w, G' \odot H') \\
&= nneg(V)^-(w, F').
\end{aligned}$$

- F is $G \to H$. By I.H., $V^-(w, G) = nneg(V)^-(w, G')$ and $V^-(w, H) = nneg(V)^-(w, H')$.

$$\begin{aligned}
V^-(h, F) &= V^-(h, G \to H) \\
&= min(\to(V^-(h, G), V^-(h, H)), \to(V^-(t, G), V^-(t, H))) \\
&= min(\to(nneg(V)^-(h, G'), nneg(V)^-(h, H')), \\
&\qquad \to(nneg(V)^-(t, G'), nneg(V)^-(t, H'))) \\
&= nneg(V)^-(h, G' \to H') \\
&= nneg(V)^-(h, F').
\end{aligned}$$

And

$$\begin{aligned}
V^-(t, F) &= V^-(t, G \to H) \\
&= \to(V^-(t, G), V^-(t, H)) \\
&= \to(nneg(V)^-(t, G'), nneg(V)^-(t, H')) \\
&= nneg(V)^-(t, F').
\end{aligned}$$

Now notice that, for any valuation V, it must be the case that for all atoms $p \in \sigma$, $V^-(w, p) \leq V^+(w, p)$, i.e. $V^-(w, p) + 1 - V^+(w, p) \leq 1$. It follows that

$$nneg(V)^-(w, p) + nneg(V)^-(w, np) \leq 1.$$

Therefore, for all atoms p,

$$\begin{aligned} nneg(V)^-(w, \neg_s(p \otimes_l np)) &= 1 - nneg(V)^-(t, (p \otimes_l np)) \\ &= 1 - \otimes_l(nneg(V)^-(t,p), nneg(V)^-(t,np)) \\ &= 1 - max(nneg(V)^-(t,p) + nneg(V)^-(t,np) - 1, 0) \\ &= 1 - 0 = 1. \end{aligned}$$

It follows that

$$\begin{aligned} V^-(w, F) &= nneg(V)^-(w, F') \\ &= \otimes_m(nneg(V)^-(w, F'), 1) \\ &= \otimes_m(nneg(V)^-(w, F'), nneg(V)^-(w, \otimes_m_{p \in \sigma} \neg_s(p \otimes_l np))) \\ &= nneg(V)^-(w, F' \otimes_m \otimes_m_{p \in \sigma} \neg_s(p \otimes_l np)) \\ &= nneg(V)^-(w, nneg(F)). \end{aligned}$$

∎

Corollary 5. *For any fuzzy formula F that may contain strong negation, a valuation V is a model of F iff $nneg(V)$ is a model of $nneg(F)$.*

Proof. By Lemma 14, $V^-(h, F) = nneg(V)^-(h, nneg(F))$. So $V^-(h, F) = 1$ iff $nneg(V)^-(h, nneg(F)) = 1$, i.e., V is a model of F iff $nneg(V)$ is a model of $nneg(F)$. ∎

Lemma 15. *for all atoms $a \in \sigma$, $V(h, a) = V(t, a)$ iff*

$$nneg(V)(h, a) = nneg(V)(t, a) \text{ and } nneg(V)(h, na) = nneg(V)(t, na).$$

Proof. For all atoms $a \in \sigma$,

$$V(h, a) = V(t, a)$$
iff $V^-(h, a) = V^-(t, a)$ and $V^+(h, a) = V^+(t, a)$
iff $nneg(V)(h, a) = nneg(V)(t, a)$ and $1 - V^+(h, a) = 1 - V^+(t, a)$
iff $nneg(V)(h, a) = nneg(V)(t, a)$ and $nneg(V)(h, na) = nneg(V)(t, na)$.

∎

Lemma 16. *For two valuations V and V_1 of signature σ, it holds that $V_1 \prec V$ iff $nneg(V_1) \prec nneg(V)$.*

Proof.

$V_1 \prec V$

iff for all atoms p (from σ), $V(t,p) = V_1(t,p)$, $V(h,p) \subseteq V_1(h,p)$ and
there exists an atom a (from σ) such that $V(h,a) \subset V_1(h,a)$

iff for all atoms p, $V^-(t,p) = V_1^-(t,p)$, $V^+(t,p) = V_1^+(t,p)$,
$\quad V_1^-(h,p) \leq V^-(h,p)$, $V_1^+(h,p) \geq V^+(h,p)$, and
there exists an atom a such that $V_1^-(h,a) < V^-(h,a)$ or $V_1^+(h,a) > V^+(h,a)$

iff for all atoms p, $V^-(t,p) = V_1^-(t,p)$, $V^+(t,p) = V_1^+(t,p)$,
$\quad V_1^-(h,p) \leq V^-(h,p)$, $V_1^+(h,p) \geq V^+(h,p)$, and
there exists an atom a such that $V_1^-(h,a) < V^-(h,a)$ or
$\quad 1 - V_1^+(h,a) < 1 - V^+(h,a)$

iff for all atoms p, $V^-(t,p) = V_1^-(t,p)$, $V^+(t,p) = V_1^+(t,p)$,
$\quad V_1^-(h,p) \leq V^-(h,p)$, $1 - V_1^+(h,p) \leq 1 - V^+(h,p)$, and
there exists an atom a such that $V_1^-(h,a) < V^-(h,a)$ or
$\quad nneg(V_1)^-(h, na) < nneg(V)^-(h, na)$

iff for all atoms p and np, $nneg(V)(t,p) = nneg(V_1)(t,p)$,
$\quad nneg(V)(t,np) = nneg(V_1)(t,np)$,
$\quad nneg(V)(h,p) \subseteq nneg(V_1)(h,p)$,
$\quad nneg(V)(h,np) \subseteq nneg(V_1)(h,np)$, and
there exists an atom a or na such that $nneg(V)(h,a) \subset nneg(V_1)(h,a)$ or
$\quad nneg(V)(h,na) \subset nneg(V_1)(h,na)$

iff $nneg(V_1) \prec nneg(V)$.

∎

Proposition 3. *For any fuzzy formula F that may contain strong negation, a valuation V is an equilibrium model of F iff $nneg(V)$ is an equilibrium model of $nneg(F)$.*

Proof. Let σ be the underlying signature of F, and let σ' be the extended signature $\sigma \cup \{np \mid p \in \sigma\}$.

($\Rightarrow$) Suppose V is an equilibrium model of F. Then $V(h,p) = V(t,p)$ for all atoms $p \in \sigma$ and V is a model of F. By Lemma 15, $nneg(V)(h,a) = nneg(V)(t,a)$ for all atoms $a \in \sigma'$, and by Corollary 5, $nneg(V)$ is a model of $nneg(F)$.

Next we show that there is no $V_1 \prec nneg(V)$ such that V_1 is a model of $nneg(F)$. Suppose, for the sake of contradiction, that there exists such V_1. We construct V' as
$$V'(w,p) = [V_1^-(w,p), 1 - V_1^-(w, np)]$$
for all atoms $p \in \sigma$. It is clear that $nneg(V') = V_1$. We will check that

(i) V' is a valid N5 valuation,

(ii) $V' \prec V$, and

(iii) V' is a model of F.

These claims contradicts the assumption that V is an equilibrium model of F. Consequently, we conclude that $nneg(V)$ is an equilibrium model of F' once we establish these claims.

To check (i), we need to show that $0 \leq V_1^-(w,p) \leq 1 - V_1^-(w,np) \leq 1$ and $V'(t,p) \subseteq V'(h,p)$ for all atoms $p \in \sigma$. Together they are equivalent to checking
$$0 \leq V_1^-(h,p) \leq V_1^-(t,p) \leq 1 - V_1^-(t,np) \leq 1 - V_1^-(h,np) \leq 1$$

The first and the last inequalities are obvious. The second and the fourth inequalities are clear from the fact that V_1 is a valid valuation. To show the third inequality, since V is a valid valuation,
$$V^-(t,p) \leq V^+(t,p)$$
which is equivalent to
$$nneg(V)^-(t,p) \leq 1 - nneg(V)^-(t,np). \tag{1}$$

Since $V_1 \prec nneg(V)$, we have $V_1(t,a) = nneg(V)(t,a)$ for all atoms $a \in \sigma'$, so (1) is equivalent to
$$V_1^-(t,p) \leq 1 - V_1^-(t,np).$$

So (i) is verified.

To check claim (ii), notice $nneg(V') \prec nneg(V)$ since $V_1 \prec nneg(V)$ and $V_1 = nneg(V')$. Then the claim follows by Lemma 16.

To check claim (iii), notice $nneg(V')$ is a model of $nneg(F)$ since V_1 is a model of $nneg(F)$, and $V_1 = nneg(V')$. Then the claim follows by Corollary 5.

($\Leftarrow$) Suppose $nneg(V)$ is an equilibrium model of $nneg(F)$. Then $nneg(V)(h,a) = nneg(V)(t,a)$ for all atoms $a \in \sigma'$ and $nneg(V)$ is a model of $nneg(F)$. By Lemma

15, $V(h,a) = V(t,a)$ for all atoms $a \in \sigma'$ and by Corollary 5, V is a model of F. Next we show that there is no $V_1 \prec V$ such that V_1 is a model of F. Suppose, for the sake of contradiction, that there exists such V_1. Then by Lemma 16, $nneg(V_1) \prec nneg(V)$ and by Corollary 5, $nneg(V_1)$ is a model of $nneg(F)$, which contradicts that $nneg(V)$ is an equilibrium model of $nneg(F)$. Consequently, we conclude that V is an equilibrium model of F. ∎

Theorem 7. *For any fuzzy formula F of signature σ that may contain strong negation,*

(a) *A valuation V of σ is a fuzzy equilibrium model of F iff $V(h,p) = V(t,p)$ for all atoms p in σ and $I_{nneg(V)}$ is a stable model of $nneg(F)$ relative to $\sigma \cup \{np \mid p \in \sigma\}$.*

(b) *An interpretation I of $\sigma \cup \{np \mid p \in \sigma\}$ is a stable model of $nneg(F)$ relative to $\sigma \cup \{np \mid p \in \sigma\}$ iff $I = I_{nneg(V)}$ for some fuzzy equilibrium model V of F.*

Proof. (a) ($\Rightarrow$) Suppose V is an equilibrium model of F. By the definition of an equilibrium model, $V(h,a) = V(t,a)$ for all atoms a. By Proposition 3, $nneg(V)$ is an equilibrium model of $nneg(F)$. By Theorem 6, $I_{nneg(V)}$ in the sense of Theorem 6 is a stable model of $nneg(F)$ relative to $\sigma \cup \{np \mid p \in \sigma\}$.

($\Leftarrow$) Suppose $I_{nneg(V)}$ in the sense of Theorem 6 is a stable model of $nneg(F)$ relative to $\sigma \cup \{np \mid p \in \sigma\}$. By Theorem 6, $nneg(V)$ is an equilibrium model of $nneg(F)$. By Proposition 3, V is an equilibrium model of F.

(b) ($\Rightarrow$) Suppose I is a stable model of $nneg(F)$ relative to $\sigma \cup \{np \mid p \in \sigma\}$. Then $I \models \bigotimes_{m \neg_s}(p \otimes_l np)$. It follows that, for all atoms $p \in \sigma$, $v_I(p) + v_I(np) \leq 1$ and
$\;p\in\sigma$
thus $v_I(p) \leq 1 - v_I(np)$. Construct the valuation V of σ by defining $V(w,p) = [v_I(p), 1 - v_I(np)]$. Clearly, I can be viewed as $I_{nneg(V)}$. By Theorem 6, $nneg(V)$ is an equilibrium model of $nneg(F)$. By Proposition 3, V is an equilibrium model of F.

($\Leftarrow$) Take any fuzzy equilibrium model V of F. By Proposition 3, $nneg(V)$ is an equilibrium model of $nneg(F)$. By Theorem 6, $I_{nneg(V)}$ is a stable model of $nneg(F)$. ∎

A.9 Proof of Theorem 8

Theorem 8. *For any fuzzy formulas F and G, I is a stable model of $F \otimes \neg G$ (relative to $\mathbf{p}$) if and only if I is a stable model of F (relative to $\mathbf{p}$) and $I \models \neg G$.*

Proof.
($\Rightarrow$) Suppose I is a 1-stable model of $F \otimes \neg G$ relative to $\mathbf{p}$. Then $I \models_1 F \otimes \neg G$ and there is no $J <^{\mathbf{p}} I$ such that $J \models_1 F^I \otimes (\neg G)^I$. By Lemma 6, $I \models_1 F$ and $I \models_1 \neg G$. Note that $v_I(\neg G) = 1$. Further, it can be seen that there is no $J <^{\mathbf{p}} I$ such that $J \models_1 F^I$, since otherwise this J must satisfy $J <^{\mathbf{p}} I$ and $J \models_1 F^I \otimes v_I(\neg G)$ (i.e., $J \models_1 F^I \otimes (\neg G)^I$), which contradicts that I is 1-stable model of $F \otimes \neg G$ relative to $\mathbf{p}$. We conclude that I is a 1-stable model of F relative to $\mathbf{p}$ and $I \models_1 \neg G$.

($\Leftarrow$) Suppose I is a 1-stable model of F relative to $\mathbf{p}$ and $I \models_1 \neg G$. Then $I \models_1 F$ and $I \models_1 \neg G$,. By Lemma 6, $I \models_1 F \otimes \neg G$.[12] Next we show that there is no $J <^{\mathbf{p}} I$ such that $J \models_1 (F \otimes \neg G)^I = F^I \otimes (\neg G)^I$. Suppose, to the contrary, that there exists such J. Then by Lemma 6, $J \models_1 F^I$. This, together with the fact $J <^{\mathbf{p}} I$, contradicts that I is a 1-stable model of F relative to $\mathbf{p}$. So there does not exist such J, and we conclude that I is a 1-stable model of $F \otimes \neg G$ relative to $\mathbf{p}$. ∎

A.10 Proof of Theorem 9

Proposition 4. *For any fuzzy interpretation I and any set $\mathbf{p}$ of fuzzy atoms, $I \models \mathbf{p}^{\text{ch}}$.*

Proof. Suppose $\mathbf{p} = (p_1, \ldots, p_n)$.

$$\begin{aligned}
v_I(\mathbf{p}^{\text{ch}}) &= v_I(\{p_1\}^{\text{ch}} \otimes \ldots \otimes \{p_n\}^{\text{ch}}) \\
&= v_I((p_1 \oplus_l \neg_s p_1) \otimes \ldots \otimes (p_n \oplus_l \neg_s p_n)) \\
&= min(v_I(p_1) + 1 - v_I(p_1), 1) \otimes \ldots \otimes min(v_I(p_n) + 1 - v_I(p_n), 1) \\
&= 1.
\end{aligned}$$

∎

Theorem 9.

(a) For any real number $y \in [0, 1]$, if I is a y-stable model of F relative to $\mathbf{p} \cup \mathbf{q}$, then I is a y-stable model of F relative to $\mathbf{p}$.

(b) I is a 1-stable model of F relative to $\mathbf{p}$ iff I is a 1-stable model of $F \otimes \mathbf{q}^{\text{ch}}$ relative to $\mathbf{p} \cup \mathbf{q}$.

Proof. (a) Suppose I is a y-stable model of F relative to $\mathbf{p} \cup \mathbf{q}$. Then clearly $I \models_y F$. Next we show that there is no $J <^{\mathbf{p}} I$ such that $J \models_y F^I$. Suppose, for the

---

[12] This does not hold if the threshold considered is not 1. For example, suppose $v_I(F) = 0.5$ and $v_I(G) = 0.5$, and consider $\otimes_l$ as the fuzzy conjunction. Clearly, $I \models_{0.5} F$ and $I \models_{0.5} G$ but $I \not\models_{0.5} F \otimes_l G$.

sake of contradiction, that there exists such J. Since $v_J(q) = v_I(q)$ for all $q \in \mathbf{q}$, J must satisfy $J <^{\mathbf{pq}} I$ and $J \models_y F^I$, which contradicts the assumption that I is a y-stable model of F relative to $\mathbf{p} \cup \mathbf{q}$. So such J does not exist, and we conclude that I is a y-stable model of F relative to $\mathbf{p}$.

(b) ($\Rightarrow$) Suppose I is a 1-stable model of F relative to $\mathbf{p}$. Clearly, $I \models F$. By Proposition 4, $I \models \mathbf{q}^{\text{ch}}$, and by Lemma 6, $I \models F \otimes \mathbf{q}^{\text{ch}}$.[13]

Next we show that there is no $J <^{\mathbf{pq}} I$ such that $J \models (F \otimes \mathbf{q}^{\text{ch}})^I$. Suppose, for the sake of contradiction, that there exists such J. Since $J \models (F \otimes \mathbf{q}^{\text{ch}})^I$, i.e.,

$$J \models F^I \otimes (q_1 \oplus_l (\neg_s q_1)^I) \otimes \ldots \otimes (q_n \oplus_l (\neg_s q_n)^I),$$

by Lemma 6, it follows that $J \models (q_k \oplus_l v_I(\neg_s q_k))$ for each $k = 1, \ldots, n$, which means that $v_J(q_k) \geq v_I(q_k)$. On the other hand, since $J <^{\mathbf{pq}} I$, we have $v_J(q_k) \leq v_I(q_k)$. So we conclude $v_J(q_k) = v_I(q_k)$.[14] Together with the assumption that $J <^{\mathbf{pq}} I$, this implies that $J <^{\mathbf{p}} I$. From $J \models (F \otimes \mathbf{q}^{\text{ch}})^I$, it follows $J \models F^I$, which contradicts that I is a stable model of F relative to $\mathbf{p}$. So such J does not exist, and we conclude that I is a stable model of F relative to $\mathbf{p} \cup \mathbf{q}$.

($\Leftarrow$) Suppose I is a 1-stable model of $F \otimes \mathbf{q}^{\text{ch}}$ relative to $\mathbf{p} \cup \mathbf{q}$. Then $I \models F \otimes \mathbf{q}^{\text{ch}}$. By Lemma 6, $I \models F$. Next we show that that there is no $J <^{\mathbf{p}} I$ such that $J \models F^I$. Suppose, for the sake of contradiction, that there exists such J.

First, it is easy to conclude $J <^{\mathbf{pq}} I$ from the fact that $J <^{\mathbf{p}} I$ since J and I agree on $\mathbf{q}$. Second, by Proposition 4, $J \models \mathbf{q}^{\text{ch}}$. Since J and I agree on $\mathbf{q}$, it follows that $J \models (\mathbf{q}^{\text{ch}})^I$. Since $J \models F^I$, by Lemma 6, $J \models F^I \otimes (\mathbf{q}^{\text{ch}})^I$, or equivalently, $J \models (F \otimes \mathbf{q}^{\text{ch}})^I$. This, together with the fact that $J <^{\mathbf{pq}} I$, contradicts that I is a 1-stable model of $F \otimes \mathbf{q}^{\text{ch}}$ relative to $\mathbf{p} \cup \mathbf{q}$. So such J does not exist, and we conclude that I is a 1-stable model of F relative to $\mathbf{p}$. ∎

---

[13] It is not necessary to have the threshold $y = 1$ for this to hold. In general, suppose $I \models_y F$. Since $I \models_1 \mathbf{p}^{\text{ch}}$, by the property of fuzzy conjunction($\otimes(x, 1) = x$), $I \models_y F \otimes \mathbf{p}^{\text{ch}}$.

[14] This cannot be concluded if we have $J \models_y (q_k \oplus_l v_I(\neg_s q_k))$, instead of $J \models_1 (q_k \oplus_l v_I(\neg_s q_k))$. So this direction of the theorem does not hold for an arbitrary threshold.

MULTI-ATTRIBUTE DECISION MAKING WITH WEIGHTED DESCRIPTION LOGICS

ERMAN ACAR, MANUEL FINK, CHRISTIAN MEILICKE, CAMILO THORNE AND HEINER STUCKENSCHMIDT
Data and Web Science Group
University of Mannheim
{erman,manuel,christian,camilo,heiner}@informatik.uni-mannheim.de

Abstract

We introduce a decision-theoretic framework based on Description Logics (DLs), which can be used to encode and solve single stage multi-attribute decision problems. In particular, we consider the background knowledge as a DL knowledge base where each attribute is represented by a concept, weighted by a utility value which is asserted by the user. This yields a compact representation of preferences over attributes. Moreover, we represent choices as knowledge base individuals, and induce a ranking via the aggregation of attributes that they satisfy. We discuss the benefits of the approach from a decision theory point of view. Furthermore, we introduce an implementation of the framework as a Protégé plugin called uDecide. The plugin takes as input an ontology as background knowledge, and returns the choices consistent with the user's (the knowledge base) preferences. We describe a use case with data from DBpedia. We also provide empirical results for its performance in the size of the ontology using the reasoner Konclude.

1 Introduction

The study of preference representation languages and decision support systems is an ongoing research subject in artificial intelligence, gaining more popularity every day. Since the inception multi-attribute utility theory (MAUT)[19], numerous approaches have been studied, including probabilistic, possibilistic, fuzzy and graphical models [9, 29, 18, 14] amongst others. One approach that has been gaining in interest over the last two decades is the use of logical languages [5, 11, 17, 20, 26, 27] to encode preferences and decision-theoretic problems.

We would like to thank all anonymous reviewers for their suggestions and comments.

In this paper, we introduce a decision-theoretic framework to encode simple single stage (or non-sequential) decision making problems using a weighted extension of Description Logics (DLs). While we presented a first sketch of the framework in [2], we now extend our formalism and give a full presentation in this paper. Furthermore, we introduce our Protégé plugin uDecide which is based on this formalism. Our framework combines ontology reasoning [4] in a basic MAUT [19] fashion to rank the available choices with respect to a set of weighted attributes that have been specified by the user; the greater the weight, the more important the attribute. This yields a compact representation for a user's preference over attributes. In that manner, it is a multi-attribute account to decision making problems. The framework may serve as a decision support system from three main perspectives: (1) in complex domains in which an extensive amount of background knowledge is required and it is hard to see the logical implications of any choice in terms of implying outcomes; (2) in scenarios which the size of the set of possible choices is too large for a human decision maker to operate; (3) when (1) and (2) are combined.

The required weighted DL (the *DL decision base*) can be built over any specific DL language. This provides flexibility in the sense that one can use tractable fragments e.g., the DL_{Lite} family [7] or $\mathcal{EL}$ [3] if scalability is important, or more expressive fragments if the domain to be modelled requires, e.g., data types, a feature desirable to express numeric domains, common in the literature of decision theory [25].

A particular feature of our approach is that we make a distinction between choices (alternatives, desired items or objects, etc.) and outcomes (which can be seen as the result of an ontological approach to decision making). In particular, an outcome is a subset of attributes which are represented by sets of description logic classes, and choices are *named individuals*. We assume that the preferences of the user are elicited (partially or completely) in the form of attributes, so that the preference relation of the user rather than talking about any specific choices e.g., a, b, talks about any anonymous or generic choices satisfying (i.e., instantiating) a given subset of attributes or criteria.

The framework and its implementation can be understood as a generic out-of-the-box expert system that turns an ontology for a specific domain into a decision support system for the domain described by that ontology. Hereby, we note that we do not make the strong claim that it can be directly applied to any decision problem scenario, but rather mean its applicability potential on domains where expert knowledge is required e.g., to select between different treatments depending on characteristics and preferences of a patient (medical domain), or a location for building a power-plant depending on the preferences of corporate management (energy industry domain). Our approach might also be used, for example, as a web-based decision support/recommender system for e-shopping. For the aforementioned perspective (1), designed to stress the importance of logic reasoning, we will give a small example in which the agent makes a decision between two cars with different specifications. For the aforementioned perspective (2), we will present a use case to illustrate, using the implemented system, uDecide, how to to generate reading proposals from thousands of choices. Loosely speaking, this use case will also demonstrate that our approach can be easily applied to any kind of domain for which an ontological representation or knowledge base is already available or can play a key role to encode domain knowledge.

In the remainder of the paper, we present the basics of (classical) utility theory and of DLs in Section 2.1. We assume that the reader has a basic familiarity with DL and omit an extensive introduction, which can be found in [4]. Next, we introduce the theoretical foundations of our framework in Section 3 and discuss it over an example. In the example, we encode a car buying agents decision making problem and show how our approach decides between two alternative cars according to the patient preferences. In Section 4, we first give a general description of our plugin. Then we present a use case which is based on an excerpt from DBpedia that deals with books and authors. This use case illustrates how to convert an ontology into an expert system by applying uDecide. Furthermore, we report on first runtime results using Konclude [31] as reasoning engine in the background. In Section 5, we discuss related works. Finally, we conclude and give a brief outline for future research in Section 6.

2 Preliminaries

2.1 Preferences and Utility

Preferences are of central importance in the study of decisions. In formal sciences such as mathematical economics, social choice theory and artificial intelligence, preferences are usually modelled as a binary relation over the set of choices $\mathcal{C}$ [16, 24, 30, 6] i.e., $c_1 \succeq c_2$, is read c_1 *is at least as good as* c_2 (for the agent which/who has the preference relation $\succ$), where $c_1, c_2 \in \mathcal{C}$. Other commonly used synonym terms are *outcome, alternative* and *object*. Moreover, it is often assumed that $\succeq$ is complete and transitive. The preference relations associated with $\succeq$ are defined as follows: for any $c, c' \in \mathcal{C}$,

$$c_1 \succ c_2 \quad \text{iff} \quad c_1 \succeq c_2 \text{ and } c_2 \not\succeq c_1, \qquad \text{(Strict preference)}$$
$$c_1 \sim c_2 \quad \text{iff} \quad c_1 \succeq c_2 \text{ and } c_2 \succeq c_1, \qquad \text{(Indifference)}$$

where the former is read c_1 *is better than* c_2, and the latter is read *the agent is indifferent between* c_1 *and* c_2. In order to represent the preference relation compactly, one introduces a *utility function* u [13], which is a function that maps a choice to a real number reflecting the degree of desire. Observe that there can be more than one such function which represents a preference relation. For the classical results that guarantees the existence of such functions, we refer to the so-called representation theorems [13]. Formally, given a choice $c \in \mathcal{C}$, a *utility function*, $u : \mathcal{C} \to \mathbb{R}$ represents $\succeq$ if

$$c_1 \succeq c_2 \quad \text{iff} \quad u(c_1) \geq u(c_2). \qquad \text{(Utility function)}$$

The associated preferences $\succ$ (strict preference) and $\sim$ (equivalent preference or indifference) are defined analogously. For instance, if it is the case that $u(\textit{low-price}) = 20$ and $u(\textit{high-price}) = 5$, this would induce the preference *low-price* $\succ$ *high-price* since $5 < 20$. Usually, choices are formalised as *values* or *elements* of *attribute*(s). Here, the choices *low-price* and *high-price* which represent any item of interest (e.g., a book, car or treatment) can be thought of as values of a single attribute *price*, or equally in set notation *price* = {*low-price, high-price*}.

Most of the time, our decisions depend on more than a single attribute; for instance, if we intend to buy a book, we are interested usually not only in its price, but also in its author, to which genre it belongs. MAUT is the extended variant of utility theory that deals with such decision problems [19, 28]. We will denote the set of attributes by $\mathcal{X}$ and refer to a specific attribute by $X_i \in \mathcal{X}$ where $i \in \{1, \ldots, |\mathcal{X}|\}$. Then, the set of choices is lifted to the cartesian product of the set of attributes that we read as (possible) outcomes, denoted Ω i.e., $\mathcal{C} \subseteq X_1 \times \ldots \times X_n = \Omega$. We say, $u : \Omega \to \mathbb{R}$ is the (multiattribute) utility function which represents $\succeq$ iff $\forall (x_1, \ldots, x_n), (y_1, \ldots, y_n) \in \Omega$,

$$(x_1, \ldots, x_n) \succeq (y_1, \ldots, y_n) \quad \text{iff} \quad u(x_1, \ldots, x_n) \geq u(y_1, \ldots, y_n).$$

Proving such representing functions exist, is a standard work in decision theory (see e.g., [13] and [19] for basic proofs).

Since the size of the Ω is large i.e., $2^{|\mathcal{X}|}$, making the assumption that u is *additive*, significantly decreases the computational complexity. An additive function satisfies the following.

$$u(x_1, \ldots, x_n) = u(x_1) + \cdots + u(x_n) \qquad \text{(Additivity)}$$

where $(x_1, \ldots, x_n) \in \Omega$.

The basic rationality principle in utility theory is that a rational agent should always try to maximise its utility, or *should take the choice with the maximum utility*:

$$Opt(\mathcal{C}) := \arg\max_{c \in \mathcal{C}} u(c) \qquad \text{(Optimal choice)}$$

where $Opt(\mathcal{C})$ corresponds to maximal elements in $\mathcal{C}$ w.r.t. u (therefore w.r.t. $\succeq$). Note that there can be more than one (since utility values of choices are not unique). The class of decision making problems which we limit ourselves with, are *discrete choice* problems [34], that is $\mathcal{C}$ is a finite set. Examples of such discrete choice problems include choosing a medical treatment program with respect to a patient's criteria, or selecting a location for a nuclear power plant following environmental and financial criteria.[1] For a comprehensive treatment on MAUT, we refer the reader to standard texts [12, 19].

2.2 Description Logics

We assume that the reader has some familiarity with DL. If that is not the case, we refer the interested reader to [4]. The framework that we are presenting is independent from the choice of a specific DL language, that is, higher the expressivity of the chosen DL language, more complex the statements one can use about the problem domain. This comes with the usual implication of the more demanding computational resource requirements. In what follows we recall briefly the basics of DL.

DL signatures can be thought of as triples (N_C, N_R, N_I), where N_C is the set of atomic concepts, N_R is the set of role names, and N_I is the set of individuals. Along the text, we assume the *unique name assumption*, which means that different individuals have different names. We denote concepts or classes by C and D, roles by r and S, and individuals as a and b.[2] Concept descriptions are defined inductively from N_C as $\neg C$, $C \sqcap D$, and $C \sqcup D$ if C and D are concept descriptions, and $\exists r.C$ and $\forall r.C$ if $r \in N_R$ and C is a concept description. The top concept $\top$ is abbreviation for $C \sqcup \neg C$ and the bottom concept $\bot$ for $\neg \top$. An interpretation is a pair $\mathcal{I} := (\Delta^\mathcal{I}, \cdot^\mathcal{I})$ where the domain $\Delta^\mathcal{I}$ is a non-empty set, and $\cdot^\mathcal{I}$ is an interpretation function which assigns to every concept name C a set $C^\mathcal{I} \subseteq \Delta^\mathcal{I}$ and to every role name R a binary relation $R^\mathcal{I} \subseteq \Delta^\mathcal{I} \times \Delta^\mathcal{I}$. It is defined inductively for every concept description as follows; $(\neg C)^\mathcal{I} := \Delta^\mathcal{I} \setminus C^\mathcal{I}$, $(C \sqcap D)^\mathcal{I} := C^\mathcal{I} \cap D^\mathcal{I}$, $(C \sqcup D)^\mathcal{I} := C^\mathcal{I} \cup D^\mathcal{I}$, $(\exists r.C)^\mathcal{I} := \{a \in \Delta^\mathcal{I} \mid \text{exists } b, (a,b) \in r^\mathcal{I} \text{ and } b \in C^\mathcal{I}\}$, and $(\forall r.C)^\mathcal{I} := \{a \in \Delta^\mathcal{I} \mid \text{for all } b, (a,b) \in r^\mathcal{I} \text{ implies } b \in C^\mathcal{I}\}$. Any other extension is defined accordingly, and will be clarified when it is necessary.

In DLs, there is a distinction between *terminological knowledge* (TBox) and *assertional knowledge* (Abox). A TBox is a set of *concept inclusions*: $C \sqsubseteq D$ which has the semantics $C^\mathcal{I} \subseteq D^\mathcal{I}$ under any interpretation $\mathcal{I}$. Furthermore, a *concept*

---

[1] This setting is orthogonal to continuous set of choices (possibly a vector of arbitrary numerical quantities) which corresponds to a real-valued optimisation problem.

[2] We will use the terms *concept* or *class* interchangeably. Both terms are common in DL literature.

definition is $C \equiv D$ if $C \sqsubseteq D$ and $C \sqsubseteq D$. An ABox is a set of concept assertions $C(a)$ where $a \in N_I$ and $C(a)^\mathcal{I} := a^\mathcal{I} \in C^\mathcal{I}$, and role assertions $R(a,b)$ where $(a,b) \in N_I \times N_I$ and $R(a,b)^\mathcal{I} := (a^\mathcal{I}, b^\mathcal{I}) \in R^\mathcal{I}$.

A concept is satisfiable if there is an interpretation $\mathcal{I}$ such that $C^\mathcal{I} \neq \emptyset$. A concept is satisfiable with respect to a TBox $\mathcal{T}$ if and only if (iff) there is a model $\mathcal{I}$ of $\mathcal{T}$ such that $C^\mathcal{I} \neq \emptyset$. An interpretation $\mathcal{I}$ satisfies a concept inclusion $C \sqsubseteq D$ ($\mathcal{I} \models C \sqsubseteq D$) iff $C^\mathcal{I} \subseteq D^\mathcal{I}$. A concept C is subsumed by a concept D with respect to a TBox $\mathcal{T}$ if $C^\mathcal{I} \subseteq D^\mathcal{I}$ for every model of $\mathcal{I}$ of $\mathcal{T}$ ($C \sqsubseteq_\mathcal{T} D$ or $\mathcal{T} \models C \sqsubseteq D$). An interpretation $\mathcal{I}$ is satisfies a TBox $\mathcal{T}$ if and only if $\mathcal{I}$ satisfies every concept inclusion in $\mathcal{T}$. A TBox $\mathcal{T}$ is called *coherent* if all of the appearing concepts in $\mathcal{T}$ are satisfiable. We say that an assertion α is entailed by an ABox $\mathcal{A}$ and write $\mathcal{A} \models \alpha$, if every model of $\mathcal{A}$ also satisfies α. An ABox $\mathcal{A}$ is consistent w.r.t. a TBox $\mathcal{T}$ if there is an interpretation $\mathcal{I}$ which satisfies $\mathcal{T}$ and $\mathcal{A}$. We call the pair $\mathcal{K} := \langle \mathcal{T}, \mathcal{A} \rangle$ a knowledge base, and say that $\mathcal{K}$ is satisfiable if $\mathcal{A}$ is consistent w.r.t. $\mathcal{T}$. One basic reasoning service we will use is *instance check*: given a knowledge base $\mathcal{K}$ and an assertion α, to check whether $\mathcal{A} \models \alpha$.

A concrete domain $\mathcal{D}$ is a pair $(\Delta^\mathcal{D}, pred(\mathcal{D}))$ where $\Delta^\mathcal{D}$ is the domain of $\mathcal{D}$ and $pred(\mathcal{D})$ is the set of predicate names of $\mathcal{D}$. It is assumed that $\Delta^\mathcal{I} \cap \Delta^\mathcal{D} = \emptyset$, and each $P \in pred(\mathcal{D})$ with arity n is associated with $P^\mathcal{D} \subseteq (\Delta^\mathcal{D})^n$. We will denote *functional roles* with lower case r. In DL with concrete domains, it is assumed that N_R is partitioned into a set of *functional roles* and the set of ordinary roles. A role r is *functional* if for every $(x,y) \in r$ and $(w,z) \in r$ it implies that $x = w \implies y = z$. Functional roles, in the extended language, are interpreted as partial functions from $\Delta^\mathcal{I}$ to $\Delta^\mathcal{I} \times \Delta^\mathcal{D}$. Functional roles and ordinary roles are both allowed to be used with both the existential quantification and the universal quantification. A concrete domain is required to be closed under negation (denoted by $\overline{P}$), in order to be able to compute the negation normal form of the concepts defined via extended constructs.

3 Weighted DLs for Decision Making

In this section, we introduce the theoretical underpinning of our plugin, which is a framework based on weighted DLs. In a loose sense, we will follow a specific-to-generic path while introducing definitions, ending up defining the generic framework, the DL *decision base*.

As an ontological approach to decision making, our aim is to use an *a priori* preference relation over attributes (ontological classes) to derive an *a posteriori* preference relation over choices (ontological individuals). To this end, we define a *priori* (given by the user) a utility function U over $\mathcal{X}$. Then we extend it to the subset of attributes, via a utility function u defined over choices using logical entailment. The use of lower case u, has two motivations: *i*) dealing with a technical subtlety, that is, a choice is an individual while a corresponding outcome is a set of classes (which are mathematically different types of objects), and *ii*) flexibility to aggregate U in different forms e.g., *max*, *mean*, or a customized arbitrary aggregation.

3.1 Modeling Attributes as DL Concepts

We represent each attribute in the original decision making problem by a class. Furthermore, for every value of an attribute in the original decision making problem, we introduce a new (sub)class to the set of classes at hand. For instance, if *colour* is an attribute in the decision making problem we would like to model, we simply represent it by the class *Colour* (i.e., $Colour \in \mathcal{X}$). Being a colour can be considered as it has a desirability on its own. Moreover, if its value is available; if *blue* is a value of the attribute *colour*, we extend our attribute set $\mathcal{X}$ simply by adding the class *Blue*, as a subclass of *Colour*. Furthermore, for each further available value e.g., *red*, we can add *Red* as a subclass, along with adding an axiom guaranteeing the disjointness e.g., $Red \sqsubseteq \neg Blue$. Note that w.l.o.g. this simple process will yield a binary term vector for $\mathcal{X}$. This is indeed our aim since in the sequel the aggregation of utilities of classes will be done with respect to the entailed class membership (i.e., $\mathcal{K} \models X(c)$) which has two possible cases in return : an individual c (as a choice) is either a member of the class X or not.

We assume a total and transitive preference relation (i.e., $\succeq_{\mathcal{X}}$) over an ordered set of attributes $\mathcal{X}$ that are not necessarily atomic, and a function $U : \mathcal{X} \rightarrow \mathbb{R}$ that represents $\succeq$ (i.e., $U(X_1) \geq U(X_2)$ iff $X_1 \succeq_{\mathcal{X}} X_2$ for $X_1, X_2 \in \mathcal{X}$). The function U can be thought of as a weight function, which assigns an *a priori* weight to each class $X \in \mathcal{X}$, and makes the description logic *weighted*. We denote **the utility of a class** $X \in \mathcal{X}$ by $U(X)$. This reflects an agent's preference relation over the set of attributes $\mathcal{X}$. The greater the utility an attribute has, the more preferable the attribute is. Furthermore, we partition the attribute set $\mathcal{X}$ into two subsets; *desirable* that is the set of attributes with non-negative weights, denoted $\mathcal{X}^+$, and *undesirable* $\mathcal{X}^-$, i.e., $X \in \mathcal{X}$ iff $U(X) \geq 0$ and $\mathcal{X} = \mathcal{X}^+ \cup \mathcal{X}^-$ with $\mathcal{X}^+ \cap \mathcal{X}^- = \emptyset$. Intuitively, any attribute that is not desirable is undesirable, and a zero-weighted attribute can be interpreted as desirable with zero utility.

3.2 From Utility of Criteria to Utility of Choices

We call N_I as the set of named individuals. A *choice* is an individual $c \in N_I$. We denote by $\mathcal{C}$ the finite set of choices. In order to derive a preference relation (*a posteriori*) over $\mathcal{C}$ (i.e., $\succeq_\mathcal{C}$) which respects $\succeq_\mathcal{X}$, we will introduce a utility function $u(c) \in \mathbb{R}$, which measures **the utility of a choice** c relative to the attribute set $\mathcal{X}$ and the utility function U over attributes as an aggregator. For simplicity, we will abuse the notation and use the symbol $\succeq$ for both choices and attributes whenever it is obvious from the context. In the following, we define a particular u, which we call σ-utility. It is intuitively defined relative to a knowledge base.

Definition 3.1 (σ-utility of a choice). *Given a consistent knowledge base $\mathcal{K}$, and a set of choices $\mathcal{C}$, the sigma utility of a choice $c \in \mathcal{C}$ is*

$$u_\sigma(c) := \sum \{U(X) \mid X \in \mathcal{X} \text{ and } \mathcal{K} \models X(c)\}.$$

It is easy to see that u_σ induces a preference relation over $\mathcal{C}$ i.e., $u_\sigma(c_1) \geq u_\sigma(c_2)$ iff $c_1 \succeq c_2$. For a representability proof, see e.g., Theorem 2.2 in [13].

Also, notice that each choice corresponds to a set of attributes, those whose membership is logically entailed e.g., $\mathcal{K} \models X(c)$. Such a summation function forms an additive multi-attribute utility function given in previous section in the sense that every choice c corresponds to an outcome.[3].

Following DL terminology and putting things together, we introduce the notion of a generic UBox, denoted by $\mathcal{U}$ which is the component we need, to generate utility functions.

Definition 3.2 (UBox). *A UBox is a pair $\mathcal{U} := (u_\sigma, U)$, where U is a utility function over $\mathcal{X}$ and u is the utility function over $\mathcal{C}$.*

Next, we introduce the key notion of *decision base*, which can be interpreted as a (formal, logical) model for an artificial agent in a decision situation, or a decision support system. A decision base is a triple which consists of a consistent *background knowledge*, a DL knowledge base $\mathcal{K}$, a finite set of available *choices* $\mathcal{C}$ which is represented as a set of individuals, a utility box, the component to encode user preferences and to generate a respective a utility function. For simplicity, we will drop the subscript σ and write u instead.

Definition 3.3 (Decision Base). *A decision base is a triple $\mathcal{D} = (\mathcal{K}, \mathcal{C}, \mathcal{U})$ where $\mathcal{K} = (\mathcal{T}, \mathcal{A})$ is a consistent knowledge base, $\mathcal{T}$ is a TBox and $\mathcal{A}$ is an ABox, $\mathcal{C} \subseteq N_I$ is the set of choices, and $\mathcal{U} = (u, U)$ is a UBox.*

---

[3]Note that such simple additive utility function is a strong simplification in decision theory literature such that it assumes implicitly that those values can be added. In contrast, additive forms are common in Artificial Intelligence literature [28].

Informally, the role of $\mathcal{K}$ is to provide assertional information about the choices at hand, along with the general terminological knowledge information that the agent may require to reason further over choices. Note that this is just as in the case of σ-utility, that is meant to measure the value of a choice with respect to the classes (possibly deduced) which it belongs to. In this work, we will restrict ourselves to u_σ and the *maximality principle* (picking up the choice(s) with the maximum utility).

The proposed ontological approach to decision making provides an intuitive relation between attributes. Assume that we limit ourselves to desirable attributes $\mathcal{X}^+$. Then *ceteris paribus* anything that belongs to a particular class should be at least as desirable as something that belongs to its superclass. For instance, a *new sport car* is at least as desirable as a *sport car* (since anything that is a *new sport car* is also a *sport car* i.e., *new sport car* $\sqsubseteq$ *sport car*). Generalizing this fact about desirable attributes: the more specific the attributes a choice satisfies, the more utility it gets. The opposite is the case for undesirable attributes: the more specific the attributes a choice satisfies, the less utility it gets. The following result formalizes this idea.

Proposition 1. *Let $\mathcal{K}$ be a knowledge base from a decision base $\mathcal{D}$ and $c_1, c_2 \in \mathcal{C}$ be any two choices. i) Assume that c_1 and c_2 are instances of exactly the same set of attributes from $\mathcal{X}^-$. If for every $X_2 \in \mathcal{X}^+$ with $\mathcal{K} \models X_2(c_2)$ there is a $X_1 \in \mathcal{X}^+$ with $\mathcal{K} \models X_1(c_1)$ such that $\mathcal{K} \models X_1 \sqsubseteq X_2$, then $c_1 \succeq c_2$. ii) If $\mathcal{X}^+$ is mutatis mutandis replaced by $\mathcal{X}^-$ in i), then $c_2 \succeq c_1$.*

Proof. **i)** Let $c_1, c_2 \in \mathcal{C}$. By assumption, the negative utilities c_1 and c_2 get are equal (i.e., $u^-(c_1) = u^-(c_2) = \sum_{\mathcal{K} \models X(c_2) \wedge X \in \mathcal{X}^-} U(X)$). Let us call this value Δ. It follows that $u(c_1) \geq u(c_2)$ if and only if $u^+(c_1) + \Delta \geq u^+(c_2) + \Delta$ if and only if $u^+(c_1) \geq u^+(c_2)$, where $u^+(c_1)$ (resp. $u^+(c_2)$) is the overall positive utility that c_1 (resp. c_2 has). Therefore, all we need to show is that $u^+(c_1) \geq u^+(c_2)$. Now assume that $u^+(c_2)$ is generated by the set $\{X_2^1, \ldots, X_2^m\} \subseteq \mathcal{X}^+$, and $u^+(c_1)$ is generated by the set $\{X_1^1, \ldots, X_1^k\} \subseteq \mathcal{X}^+$. Since for every $X_2 \in \mathcal{X}^+$ with $\mathcal{K} \models X_2(c_2)$ there is a $X_1 \in \mathcal{X}^+$ with $\mathcal{K} \models X_1(c_1)$ such that $\mathcal{K} \models X_1 \sqsubseteq X_2$ (by assumption), it follows that $\{X_2^1, \ldots, X_2^m\} \subseteq \{X_1^1, \ldots, X_1^k\}$, hence $u^+(c_1) \geq u^+(c_2)$. **ii)** Similar to the previous case. □

The following corollary is a natural result of this approach to decision theory, which says that two choices are of same desirability (i.e., indistinguishable w.r.t. desirability) if they belong to exactly the same classes.

Corollary 1 (Indistinguishableness). *Let $\mathcal{D}$ be a decision base, then for any $c_1, c_2 \in \mathcal{C}$, $c_1 \sim c_2$ iff $\{X_1 \in \mathcal{X} \mid \mathcal{K} \models X_1(c_1)\} = \{X_2 \in \mathcal{X} \mid \mathcal{K} \models X_2(c_2)\}$.*

Proof. By applying Proposition 1 in both directions. □

The intuitive explanation for Corollary 1 is that we measure the desirability (and non-desirability) of things, according to what they are, or which classes they do belong to. This brings forward the importance of reasoning, since it might not be obvious at all that two choices actually belong to exactly the same classes as attributes.

Remark 1. Note that w.l.o.g. the complexity of calculating the optimal choice is as high as the complexity of the instance checking problem in the employed DL language, since to calculate the optimal choice, given m choices and n attributes, in the worst case there are $m \cdot n$ instance checking to be performed.

Decision bases offer high flexibility in representing preferences, due to having both qualitative (logical) and quantitative (weights) components. The following example illustrates their main properties.

3.3 Example: Car Buyer

Consider an agent who wants to buy a second hand sports car. After visiting various car dealers, he finds two alternatives as fair deals; a sport Mazda (*Mx-5 Miata Roadster, 2013*) which fits his original purpose and a BMW (*335i Sedan, 2008*) which is also worth considering since it has a very strong engine (*300 horsepower (hp)*) and also comes with a sport kit. The car buyer's decision base (background knowledge $(\mathcal{T}, \mathcal{A})$, choices $\mathcal{C} = \{car_1, car_2\}$, and attributes mentioned in $\mathcal{U}$) is as in Figure 1.

As the use of numerical domains is common to classical Decision Theory, we will use the language with concrete domains. If the reader is already familiar with concrete domains, she can skip the technical definitions and go directly to Figure 1.

Let us clarify concrete domains and predicates which are used in the example. We take the concrete domain Car and $\Delta^{Car} := \Delta^\$ \cup \Delta^{sec} \cup \Delta^{mpg} \cup \Delta^{mph} \cup \Delta^{hp}$ with $\Delta^\$ \cap \Delta^{sec} \cap \Delta^{mpg} \cap \Delta^{mph} \cap \Delta^{hp} = \emptyset$, and $pred(Car) := pred(\$) \cup pred(mpg) \cup pred(mph) \cup pred(sec)$. For $\mathbb{S} \subset \mathbb{Q}$ is a sufficiently large finite set of rationals, we define the partition (of the domain Δ^{Car}) $\Delta^\$$ as $\{i\$ \mid i \in \mathbb{S}\}$, $pred(\$) := \{<_\$, >_\$, \geq_\$, \leq_\$, =_\$, \neq_\$\}$. $(<_\$)^\$(x, y) = \{(x, y) \in \Delta^\$ \times \Delta^\$ \mid i, j \in \mathbb{S}$ with $x := i\$$ and $y := j\$$ such that $i < j\}$. Other predicates are defined similarly in an obvious way parallel to usual binary relations over $\mathbb{S}$. For convenience, we extend $pred(\$)$ with finitely many unary predicates in the form of $<_x := \{\forall y \in \Delta^\$ \mid <_\$ (x, y)\}$ and also of $>_x, \leq_x, \geq_x, =_x, \neq_x$ which are similarly defined, enough to express the intended TBox. Note that $pred(\$)$ is closed under negation: $\overline{<_\$}(x, y) =\geq_\$ (x, y)$, etc. For other partitions, we take $\Delta^{sec} := \{i\ sec \mid i \in \mathbb{S}^+ \setminus \{0\}\}$, $\Delta^{mpg} := \{i\ mpg \mid i \in \mathbb{S}\}$, $\Delta^{mph} := \{i\ mph \mid i \in \mathbb{S}\}$, $\Delta^{hp} := \{i\ hp \mid i \in \mathbb{S} \setminus \{0\}\}$. The rest of the respective predicate names and functional roles are defined in an obvious way ($hasPrice^\mathcal{I} : \Delta^\mathcal{I} \times \Delta^\$$, $hasKit : \Delta^\mathcal{I} \times \Delta^\mathcal{I}$, etc).

$\mathcal{T} = \{\exists hasPrice. \leq_{30000\ \$} \equiv InexpensiveCar,$
$\quad ExpensiveCar \sqsubseteq HighClassCar,$
$\quad HighClassCar \sqsubseteq PrestigiousCar,$
$\quad \exists hasFuelConsumpt. \geq_{20mpg} \equiv EconomicCar,$
$\quad Roadster \sqsubseteq PrestigiousCar,$
$\quad MiddleClassCar \sqcap HighClassCar \sqsubseteq \bot,$
$\quad SportsCar \sqcup \exists hasHP. \geq_{200hp} \sqsubseteq StrongCar,$
$\quad 2Doors \sqcap 4Doors \sqsubseteq \bot,$
$\quad \exists has0-60mph. \leq_{7.0sec} \sqcap$
$\quad \exists hasHP. \geq_{270hp} \sqsubseteq VeryStrongCar,$
$\quad 2Doors \sqcap \neg Convertible \equiv Coupé,$
$\quad \neg\ Coupé \sqcap \neg Convertible \sqcap \neg Hatchback \sqsubseteq Sedan,$
$\quad 2Doors \sqcap \exists has0-60mph. \leq_{7.0sec} \sqcap$
$\quad \forall hasKit.SportKit \sqsubseteq SportsCar,$

$\quad Bmw \sqcap Mazda \sqsubseteq \bot,$
$\quad Bmw335i \sqsubseteq Bmw,$
$\quad \exists hasModelYear. \geq_{2012} \equiv NewCar,$
$\quad Bmw \sqsubseteq PrestigiousCar,$
$\quad SportsCar \sqcap Convertible \equiv Roadster,$
$\quad ClassicalKit \sqsubseteq Kit,$
$\quad SportKit \sqsubseteq Kit,$
$\quad Car \sqcap Kit \sqsubseteq \bot,$
$\quad ClassicalKit \sqcap SportKit \sqsubseteq \bot\}$

$\mathcal{A} = \{MazdaMx5Miata(car_1),$
$\quad hasHP(car_1, 167hp),$
$\quad hasFuelConsumption(car_1, 24mpg),$
$\quad hasModelYear(car_1, 2013),$
$\quad has0-60mph(car_1, 6.9sec),$
$\quad hasPrice(car_1, 29960\$),$
$\quad Convertible(car_1),$
$\quad 2Doors(car_1),$
$\quad ClassicalKit(kit_1)$
$\quad hasKit(car_1, kit_1),$

$\quad Bmw335i(car_2),$
$\quad hasHP(car_2, 300hp),$
$\quad hasFuelConsumption(car_2, 19mpg),$
$\quad hasModelYear(car_2, 2008),$
$\quad has0-60mph(car_2, 4, 8sec),$
$\quad hasPrice(car_2, 42560\$),$
$\quad Sedan(car_2),$
$\quad 4doors(car_2),$
$\quad SportKit(kit_2),$
$\quad hasKit(car_2, kit_2)\}$

$U = \{(InexpensiveCar, 30),$
$\quad (PrestigiousCar, 55),$
$\quad (VeryStrongCar, 50),$
$\quad (StrongCar, 40),$
$\quad (EconomicCar, 30),$
$\quad (NewCar, 35),$
$\quad (Convertible, 10),$
$\quad (Sedan, 5),$
$\quad (\exists hasKit.SportKit, 20),$
$\quad (\exists hasKit.ClasicalKit, 10)\}$

$\mathcal{C} = \{car_1, car_2\}$

Figure 1: The car buyer's background knowledge $\mathcal{K} = (\mathcal{T}, \mathcal{A})$, the set of choices $\mathcal{C} = \{car_1, car_2\}$, and preferences encoded as U. We omit the trivial axioms with the super concept Car: $HighClassCar \sqsubseteq Car$, $PrestigiousCar \sqsubseteq Car, \ldots$, etc.

According to the agent, taking $\mathcal{T}$ into account, a *Bmw* is a *prestigious car*. Considering a *200 hp* or above is enough to refer to a car as strong. An economic car should go for more than *20 miles per gallon* (*mpg*). A car is *new* if it was manufactured in *2012* or later.

Considering U in Figure 1, the agent (car buyer) is more interested in having a *prestigious car* than having an *inexpensive car*. He prefers *convertible* to *sedan*. However, these are not as important as a car to be an *economic car*, or a *strong car*. Using the given decision base, we can calculate the utility of each choice ($u_\sigma(car_1) = 220$, $u_\sigma(car_2) = 170$), which implies (by the assumption: the greater the utility, the more desirable is the choice) that $car_1 \succ car_2$.

Remark 2. Note that the DL we have used in this example is quite expressive. The aim here is to simply show that using an enough expressive DL, one can *talk* about some *natural* decision problems in a reasonable way. As an example we chose a common human decision problem, dealing with it in an automated manner to support the intuition that it can be used in, say, e-shopping scenario. For convenience, in the example we set the concrete domain to a sufficiently large finite subset of rationals. Moreover without arithmetic functions over the concrete domain, it should be convenient that it is decidable [22].

4 System Description

We implemented our approach as Protégé plugin available at the link https://code.google.com/p/udecide/. We first briefly describe the functionality and architecture of our plugin in Section 4.1. Then we present a use case that shows how a user interacts with the plugin in Section 4.2. This use case also illustrates how the plugin can be used as an out-of-the-box expert system for any knowledge domain available as ontology.

4.1 Implementation

Our Protégé plugin is compatible with both Protégé Desktop version 4.3 and 5.0. As reasoning component we used the Konclude reasoner [31] which turned out to be the best OWL reasoner for our purpose with regards to performance issues.[4] When we start Protégé Konclude is required to be running in the background. The connection to Konclude is established via OWLlink.

Our implementation is straightforward. First, an ontology needs to be loaded via the standard Protégé file menu. This ontology acts as a knowledge base $\mathcal{K}$. After

---

[4] We would like to thank Andreas Steigmiller for his support related to using Konclude.

switching to the uDecide tab, the user can specify the set of possible choices by specifying a class C defined in $\mathcal{K}$ [5]. All instances of C are treated as choices, which corresponds to the set of choices $\mathcal{C}$ in the theoretical framework. The attributes and their utility can then be specified on top of the vocabulary defined in $\mathcal{K}$. Once the type of choices and the attributes with their corresponding utility have been specified, a connection to Konclude is established via HTTP. We will illustrate these steps in the subsequent section in more details. For each attribute, we request from the reasoner all named individuals satisfying the intersection of the attribute's class expression and the class that defines the type of choices. The result shown consists of a ranked list of all individuals returned by at least one query and their utility which is derived from the satisfied attributes. Since Konclude does not support instance satisfaction queries for anonymous class expressions, we create a temporary ontology that is transferred to Konclude at runtime. We add to this ontology an equivalent classes axiom between each utility assertion's class expression and a named dummy class. We then separately query the individuals for each named dummy class.

As described on our homepage we recommend to configure Konclude to load the knowledge base already on start-up to speed up the calculation. Because of increasing computation time and memory limitations it is required to do this when working with large knowledge bases. If the knowledge base was already loaded into Konclude at start-up, only a (very small) temporary ontology will be built and used by Konclude. Otherwise the union of both the temporary ontology and the (potentially very large) knowledge base will be loaded in primary memory. This behaviour can be controlled by a checkbox which is used to specify if the knowledge base is already loaded into Konclude at start-up.

4.2 Use Case

As an illustrating use case, we applied our approach to the domain of books and authors. In particular, we used our framework to support a user in finding interesting books or authors by specifying his interests as attributes. Instead of working with an artificial example, we used an existing subset of DBpedia that deals with the chosen topic. The core domain contains relevant information about books and their authors. With respect to our use case, DBpedia suffers a bit from its restricted set of terminological axioms and its incompleteness regarding the sparse usage of some properties.

In order to illustrate how to overcome such problems, we decided to extend it with information about cities. In particular, we added for each city the country in which it is located. Furthermore, we added some axioms specifying nationality

---

[5]Recall that the number of named instances for any class is finite.

classes e.g., "Spanish" is defined as "the set of those persons that were born in or died in a city located in Spain or whose nationality is Spain, Spanish_language or Spanish_people". Thus, by using the nationality classes, the nationality of authors for whom no nationality object property assertion exists, can still be inferred by their birth and death places. We did so because in the core domain the nationality was specified directly for only 21,3% authors, while 46,9% have a "derived nationality" via our axiomatisation. Note that these axioms are not always true, because there obviously exist some people that were born in Spain whose nationality is not Spanish.

This extension illustrates that, in the context of a reasoning based approach, it is possible to leverage background knowledge that seemed not to be relevant at first sight. Information that Barcelona is located in Spain and that some author was born in Barcelona can affect the ranking of choices, if we specified that we prefer Spanish authors as an attribute. It shows that a reasoning based approach can help to overcome some problems related to incomplete data in the knowledge base. The dataset and some instructions on how to use it can be found at https://code.google.com/p/udecide/wiki/BookUseCaseExample.

Suppose that a user wants to find a new author who writes books that are similar to the ones that she likes. First of all, feasible choices have to be defined as the instances of the class *dbp:Author*. Figure 2 depicts a screenshot of the uDecide tab. The class to which the choices belong has been specified in the respective text field in the upper right corner. An arbitrary concept description can be specified as long as it is in the signature of the previously loaded ontology.

Now suppose that our user likes the authors Stephen King, Edgar Allan Poe, and H.P. Lovecraft. Thus, she adds for each of the three authors an attribute to U:

$$U = \{(\exists influencedBy.\{Stephen\_King\}, 70), \ldots\}.$$

The resulting list of attributes can be seen in the uDecide tab on the left side of Figure 2. Note that the concept descriptions are specified in the Manchester syntax[6] supported by the Protégé Editor. All attributes are specified within a dialog box that uses the auto-complete functionality of Protégé as well as its syntax checking capability. Only if a class expression is syntactically correct, a button will be enabled to add it to the UBox.

Overall, nine attributes have been specified. The first three attributes express that the user prefers authors that are influenced by her favourite authors. By adding a negative value to the fourth attribute, the user ensures that the three authors that she already knows will be ranked low in the ranking of choices. The fifth attribute is added to increase the utility of those authors that received some award by 50.

---

[6] http://www.w3.org/TR/owl2-manchester-syntax/#The_Grammar

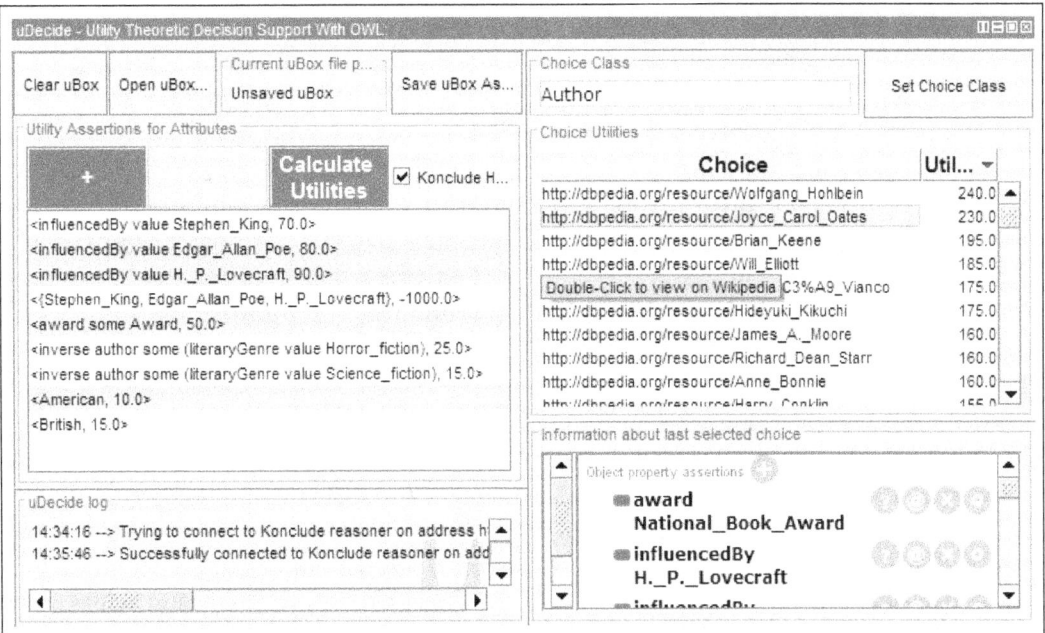

Figure 2: Screenshot of uDecide displaying a ranked list of authors according to the attribute specification of a user.

Moreover, the user specifies that she likes authors writing books that belong to the genre of horror fiction or science fiction. These attributes have a relatively low utility value. Finally, it is specified that the user likes American and British authors, slightly preferring British.

The results that are finally calculated will only include individuals that satisfy the choice class expression and at least one of the attributes.[7] This calculation is started by clicking on the "Calculate Utilities" button. The ranked choices are presented on the right side of Figure 2 in descending order based on their utility. The best choice is the author Wolfgang Hohlbein (240), followed by Joyce Carol Oates (230) and many lower ranked choices. Thus, the most reasonable choice for the user is to look at the author Wolfgang Hohlbein in more details, given that his attribute specification and the underlying knowledge base is complete and correct.

However, it might often be the case that a user wants to explore the results in

---

[7]The rationale with *at least one of the attributes* is the large knowledge bases; when one deals with large knowledge bases, most of the time, we experienced that the list is filled with so many choices with 0 utility. In order to overcome that, we filter them out by default. However it is still easy to have the whole list of choices, simply by entering the restriction class as an attribute with 0 weight.

more detail, for example to get an explanation about their ranking position. This can be done by clicking on one of the proposed choices. Figure 2 illustrates this for Joyce Carol Oates. The utility score of 230 is based on the fact that Joyce Carol Oates was influenced by Edgar Allen Poe and by H.P. Lovecraft, that she won at least one award and that she was born in New York, therefore being classified as American. Each of the satisfied attributes is highlighted in the left panel. Furthermore, all assertions about the selected choice are shown in a panel in the lower right corner. Again, we have used the Protégé default way of presenting this information. Vice versa, it is also possible to select one of the (or multiple) attributes. This results in those choices being highlighted that satisfy the selected (all the selected) attribute(s) (not shown in Figure 2).

Our use case and the presented example illustrate both the benefits as well as some drawbacks of our approach. First of all, we could apply our Protégé plugin to the domain of books and authors without the plugin requiring any further modifications or extensions. This resulted in an expert system which makes proposals about interesting authors or books. The only required ingredient was an ontology that covers the domain in an appropriate way. We decided to use DBpedia, which features a comprehensive ABox but a flat and inexpressive TBox. Thus, the potential reasoning capabilities of our approach had only a limited impact: whether a choice satisfies an attribute can be decided by a direct look-up over most attributes. This changes with the use of an expressive TBox where some of the attributes are satisfied due to a chain of non-trivial logical dependencies and background knowledge. In such a setting our approach can be used to provide non-immediately obvious recommendation, as well as elicit their *explanation*, a key issue in decision support.

4.3 Scalability

In some preliminary experiments we have also tested the scalability of our approach. We used both (1) standard datasets of our DBpedia book use case as well as (2) extended datasets (in the sense that new –not present in DBpedia– terminological and assertional knowledge was added). For each case we created subsets that differed with respect to the contained number of instances. The smallest dataset contained 13307 individuals and 51149 assertions and axioms. The largest dataset contained 54018 individuals and 300653 assertions and axioms covering all authors and books in DBpedia. The extended datasets are slightly larger in each case because they contain assertions about the countries of the cities. To further highlight the differences between the standard and the extended datasets, we used different UBoxes to measure their runtime. The UBoxes as well as the resulting choice ranking for the complete dataset can be seen in Figures 3 and 4.

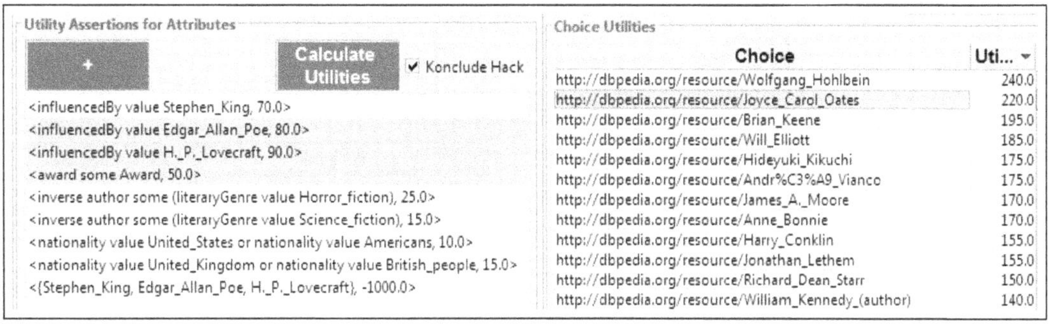

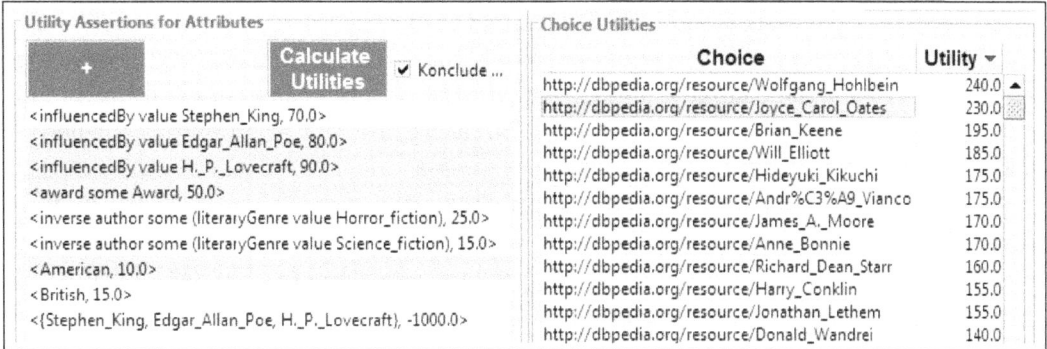

Figure 3: Screenshot of uDecide performance test displaying a ranked list of authors using the *standard* DBpedia book knowledge base.

Figure 4: Screenshot of uDecide performance test displaying a ranked list of authors using the *extended* DBpedia book knowledge base.

The UBoxes share most of their attributes, differing only on the attributes giving preference to authors with American and British nationality. In the standard case they were expressed only with the nationality object property on the domain of *dbp:Author* while in the extended case the added nationality classes were used. The results of our experiments are shown in Tables 1 and 2.

We conducted our experiments on a Win 7 desktop machine with four 3,4 Ghz cores and 8 GB DDR3-Ram using Protege 5.0 Desktop Beta build 17 and Konclude-v0.6.1-527-Windows-x64-VS05-Dynamic-Qt4.8.5. As described above, we forced Konclude to load the knowledge base already on start-up. This required less than 4 seconds for each of the datasets listed in the table. It can be seen that the runtime behaviour of uDecide is linear with respect to the size of the knowledge base. However, the runtime of uDecide is mainly based on the reasoning system that is used in the background which was Konclude in our experiments. It can also be observed that the extended case takes more time than the standard case. This

Size	# Individuals	# Axioms	Runtime
66 MB	53991	297926	5.8 sec
55 MB	47311	249126	4.6 sec
44 MB	40044	199898	4.1 sec
33 MB	33253	150900	3.1 sec
22 MB	23456	101162	2.1 sec
11 MB	13307	51149	1.0 sec

Size	# Individuals	# Axioms	Runtime
66 MB	54018	300653	8.2 sec
55 MB	47338	251454	6.9 sec
44 MB	40068	201883	5.3 sec
33 MB	32114	152023	3.8 sec
22 MB	23487	102357	2.6 sec
11 MB	13335	51893	1.3 sec

Table 2: Runtime results for different sizes of the DBpedia extended book knowledge base.

is no surprise, given that the added nationality classes, being rather complex DL concepts, required more expressive reasoning than the original *flat* attributes. In any case, even for the complete and extended dataset, we measured a runtime of not more than 8.2 seconds, which is probably still acceptable for an application where a user is waiting for an on-the-fly response.

As already described above, our experiments are currently based on an the DBpedia ontology which corresponds to a very light weight DL; without disjunction and negation, yet with (simple) concrete domains. Further experiments have to be conducted where we use more expressive knowledge bases. We are not expecting a linear runtime behaviour in such a setting.

5 Related Work

On the theory side, regarding preference representations, approaches [11, 20, 33] among many others have been proposed using propositional logic. This body of works is especially relevant to mention as they provided the inspirational basis for the design of our framework.

In [11, 33], authors investigate expressiveness of weighted propositional logics on representing classes of utility functions. In doing so, they present some correspondences between particular types of weighted formulas and well-known classes of utility functions such as monotonic, concave, and study succinctness of different types of weighted formulas for representing the same class of utility functions. The work [20] using the similar framework focuses on representing group preferences. They consider to aggregation functions sum (like our u_σ) and max (simply considering the maximum weight).

In [15] authors propose a propositional modal logic, multi-attribute preference logic, to represent and reason about multi-attribute preferences. Their main objective is to represent well-known preference relations, rather than proposing a decision support model. It must also be noted that their work is purely qualitative in contrast to ours.

On DL side, there have also been some works [26, 27] which have a machinery similar to ours. In their work, the utility of a concept (proposal) is defined as the sum of the weights of its superconcepts. In particular, in [26], Ragone et al. show how to represent preferences using weighted DL- formulas. Claiming that the definition of utility by subsumption yields unintuitive results, they base their modified definition of utility on semantic implication. This means that the utility of a concept C w.r.t. a TBox is defined as the sum of the weights of the concepts that are logically implied by C. According to terminology they used, our approach can be understood as an implication-based approach. However, they define logical implication in terms of membership, i.e., $m \models C$ iff $m \in C^{\mathcal{I}}$. The minimal model that they introduced in order to define the minimal utility value is more restrictive than ordinary models in DL. They change this definition to ordinary models in their next paper [27], while keeping the formal machinery the same (except the way they compute utilities). We should note that their preference set, which is a set of weighted concepts, is similar to our UBox. Hence, the main difference of our approach is the formal extension to multiple alternatives and the use of individuals; they represent choices as concepts. Moreover, we provide an implementation of our framework as a decision support tool.

In [32], authors show how to encode fuzzy multicriteria decision making (MCDM) problems in the formalism of fuzzy DL. They base their work on a standard feature of MCDM on continuous domains: a *decision matrix* wherein the performance score of each choice over each criteria is explicitly stated. Criteria are expressed as fuzzy concepts. The optimal choice (w.r.t the fuzzy knowledge base) is the one with the highest *maximum satisfiability degree*. The authors do not explicitly make a distinction between the knowledge base and the set of criteria. In general, the focus of the work is to show the potential and flexibility of fuzzy DL in encompassing the usual numerical methods used in MCDM, rather than leveraging a formal concept hierarchy in MCDM for expressing relations and handling inconsistencies between criteria, choices, and the knowledge base.

On the practical side, we briefly mention [8, 10] mainly because they are decision support systems and they use Protégé (to our knowledge the only ones). It is important to note that these systems have no multiattribute character. They are designed to serve solely as a clinical decision support system. The basic idea is to represent *clinical practice guidelines* (CPGs) as ontologies and make use of ontological inference, SWRL[8] and Jess[9].

6 Conclusion and Further Work

We have presented a DL based decision theoretic framework for representing basic decision problems which arise in a multiattribute fashion. This work can be understood as the first step of our working line. A preliminary version of this framework was first described in [2]. Within this paper we have improved this framework and presented a refined version. However, the main contribution of this paper is the Protégé plugin uDecide. This plugin is a straightforward implementation of the proposed framework. It computes the utility for each of the defined choices by aggregating the utility value for each satisfied attribute. Since each attribute corresponds to a class description, standard reasoning techniques can be used to check whether an attribute is satisfied. We have used the Konclude [31] reasoning system to conduct the required reasoning tasks. The results of this computation are presented to the user as a ranked list of choices.

Since we implemented our approach as a Protégé plugin, our approach can easily be used by the DL community. The current form of uDecide works with the u_σ utility function and we are currently working on extending the plug-in as an implementation of a generic decision base, where one can also define arbitrary utility functions.

We have demonstrated within our use case, where we used a DBpedia fragment concerned with the book domain, how to use uDecide as an expert system that recommends new authors to a user. Moreover, we have also shown that our current implementation, by using the reasoning system Konclude, is capable to deal with large real-world datasets. We already pointed out that the benefits of reasoning are rather limited in the context of the DBpedia subset we used. For that reason, we have to identify another use case where we can clearly show that reasoning is beneficial by making logical dependencies explicit in calculating the final ranking. It will also be interesting to perform runtime experiments on the dataset of such a use case. We are currently investigating datasets from the biomedical domain and from the domain of life sciences. A remaining task is to conclude user-experiments to check how satisfying is the choices uDecide provides for the users. This is in our agenda.

As in many other disciplines, in decision theory it is common to deal with decisions where uncertainty is present. For that reason, one major future research direction is to extend the framework with probabilistic description logics, e.g., [21, 23].

---

[8] http://www.w3.org/Submission/SWRL/
[9] http://herzberg.ca.sandia.gov/jess/

A first attempt in that direction can be found in [1]. A probabilistic approach would allow us to face the challenge of representing typical problems defined in the decision theoretic literature, along with lots of new application possibilities. In particular, a probabilistic extension would allow us to compute the expected utility of choices (lotteries) in terms of logical entailment, beside the possibility for employing different logics for different types of probabilities (e.g., subjective, statistical) that the decision problem involves. Having said that, the work we presented in this paper must be understood as a very first step in that direction with strong simplifications regarding decision problems.

A second major plan is to keep developing the theory and extend the framework with a functionality that one can use concrete domains as a part of the utility function. This would allow in return to construct ontologies which represent optimisation problems with continuous domain that is common to MAUT literature [12, 19, 18].

Another future plan is to extend the framework to sequential decisions (e.g., $\mathcal{D}_i \rightarrow \mathcal{D}_{i+1}$, sequence of decision bases). Once sequential decisions are defined, we can deal with policies, strategies and decision-theoretic planning. Besides all we believe that, with a proper multi-agent extension, uDecide could also be used to reason for group decisions, along with potential applications in computational social choice or algorithmic game theory.

References

[1] E. Acar. Computing subjective expected utility using probabilistic description logics. In U. Endriss and J. Leite, editors, *Proceedings of STAIRS 2014: the Seventh European Starting AI Researcher Symposium*, volume 264 of *Frontiers in Artificial Intelligence and Applications*, pages 21–30. IOS Press, 2014.

[2] E. Acar and C. Meilicke. Multi-attribute decision making using weighted description logics. In T. Lukasiewicz, R. Peñaloza, and A.-Y. Turhan, editors, *Proceedings of PRUV 2014: the First Workshop on Logics for Reasoning about Preferences, Uncertainty, and Vagueness*, volume 1205 of *CEUR Workshop Proceedings*, pages 1–14. CEUR-WS.org, 2014.

[3] F. Baader, S. Brandt, and C. Lutz. Pushing the EL envelope. In *Proceedings of IJCAI-05: the Nineteenth International Joint Conference on Artificial Intelligence,*, pages 364–369. Morgan Kaufmann Publishers, 2005.

[4] F. Baader, D. Calvanese, D. L. McGuinness, D. Nardi, and P. F. Patel-Schneider, editors. *The Description Logic Handbook: Theory, Implementation, and Applications*. Cambridge University Press, 2003.

[5] C. Boutilier. Toward a logic for qualitative decision theory. In *Proceedings of KR-94: the Fourth International Conference on Principles and Knowledge Representation and Reasoning*. Morgan Kaufmann Publishers, 1994.

[6] S. J. Brams and P. C. Fishburn. *Approval voting (2. ed.)*. Springer, 2007.

[7] D. Calvanese, G. D. Giacomo, D. Lembo, M. Lenzerini, and R. Rosati. Tractable reasoning and efficient query answering in description logics: The DL-lite family. *Journal of Automated Reasoning*, 39(3):385–429, 2007.

[8] M. Ceccarelli, A. Donatiello, and D. Vitale. KON^3: A clinical decision support system, in oncology environment, based on knowledge management. In *Proceedings of ICTAI-08: the Twentieth International Conference on Tools with Artificial Intelligence*, pages 206–210. IEEE Computer Society, 2008.

[9] U. Chajewska, D. Koller, and R. Parr. Making rational decisions using adaptive utility elicitation. In *Proceedings of (AAAI-00) and (IAAI-00): the Seventh Conference on Artificial Intelligence and of the Twelfth Conference on Innovative Applications of Artificial Intelligence*, pages 363–369, Menlo Park, CA, July 30– 3 2000. AAAI Press.

[10] C. Chen, K. Chen, C.-Y. Hsu, and Y.-C. J. Li. Developing guideline-based decision support systems using protégé and jess. *Computer Methods and Programs in Biomedicine*, 102(3):288–294, 2011.

[11] Y. Chevaleyre, U. Endriss, and J. Lang. Expressive power of weighted propositional formulas for cardinal preference modeling. In *Proceedings of KR-06: the International Conference on Principles of Knowledge Representation and Reasoning*, 2006.

[12] J. S. Dyer. *Multiple Criteria Decision Analysis: State of the Art Surveys*, chapter Maut — Multiattribute Utility Theory, pages 265–292. Springer New York, New York, NY, 2005.

[13] P. C. Fishburn. *Utility Theory for Decision Making*. Robert E. Krieger Publishing Co., Huntington, New York, 1969.

[14] P. H. Giang and P. P. Shenoy. Two axiomatic approaches to decision making using possibility theory. *European Journal of Operational Research*, 162(2):450–467, 2005.

[15] K. V. Hindriks, W. Visser, and C. M. Jonker. Multi-attribute preference logic. In N. Desai, A. Liu, and M. Winikoff, editors, *Proceedings of PRIMA 2010: the Thirteenth International Conference on Principles and Practice of Multi-Agent Systems: Revised Selected Papers*, pages 181–195. Springer Berlin Heidelberg, 2012.

[16] S. Kaci. *Working with Preferences: Less is More*. Springer Verlag, Berlin, 2011.

[17] S. Kaci and L. van der Torre. Reasoning with various kinds of preferences: logic, non-monotonicity, and algorithms. *Annals of Operation Research*, 163(1):89–114, 2008.

[18] C. Kahraman. *Multi-Criteria Decision Making: Theory and Applications with Recent Developments*. Springer, 2008.

[19] R. Keeney and H. Raiffa. *Decisions with Multiple Objectives: Preferences and Value Tradeoffs*. J. Wiley, New York, 1976.

[20] C. Lafage and J. Lang. Logical representation of preferences for group decision making. In *Proceedings of KR-00: the Seventh International Conference on Principles of*

Knowledge Representation and Reasoning, San Francisco, 2000.

[21] T. Lukasiewicz. Expressive probabilistic description logics. *Artificial Intelligence*, 172(6–7):852–883, 2008.

[22] C. Lutz. Description logics with concrete domains—a survey. In *Advances in Modal Logic 2002 (AiML 2002)*, Toulouse, France, 2002. Final version appeared in Advanced in Modal Logic Volume 4, 2003.

[23] C. Lutz and L. Schröder. Probabilistic description logics for subjective uncertainty. In *Proceedings of KR-10: the Twelfth International Conference on the Principles of Knowledge Representation and Reasoning*. AAAI Press, 2010.

[24] A. Mas-Colell, M. D. Whinston, and J. R. Green. *Microeconomic Theory*. Oxford University Press, New York, 1995.

[25] G. Parmigiani and L. Y. T. Inoue. *Decision Theory Principles and Approaches*. John Wiley & Sons, Ltd, 2009.

[26] A. Ragone, T. D. Noia, F. M. Donini, E. D. Sciascio, and M. P. Wellman. Computing utility from weighted description logic preference formulas. In *Post-workshop Proceedings of Declerative Agent Languages and Technologies DALT*, 2009.

[27] A. Ragone, T. D. Noia, F. M. Donini, E. D. Sciascio, and M. P. Wellman. Weighted description logics preference formulas for multiattribute negotiation. In *Proceedings of SUM-09: Third International Conference on Scalable Uncertainty Management*, 2009.

[28] S. J. Russell and P. Norvig. *Artificial Intelligence: A Modern Approach*. Pearson Education, 2 edition, 2003.

[29] Y. Shoham. Conditional utility, utility independence, and utility networks. In *Proceedings of UAI-97: the Thirteenth Conference on Uncertainty in Artificial Intelligence*, 1997.

[30] Y. Shoham and K. Leyton-Brown. *Multiagent Systems: Algorithmic, Game-Theoretic, and Logical Foundations*. Cambridge University Press, New York, NY, USA, 2008.

[31] A. Steigmiller, T. Liebig, and B. Glimm. Konclude: system description. *Journal of Web Semantics: Science, Services and Agents on the World Wide Web*, 27:78–85, 2014.

[32] U. Straccia. Multi criteria decision making in fuzzy description logics: A first step. In *Proceedings of KES, the Thirteenth International Conference on Knowledge-Based and Intelligent Information and Engineering Systems - Part 1*, 2009.

[33] J. Uckelman, Y. Chevaleyre, U. Endriss, and J. Lang. Representing utility functions via weighted goals. *Mathematical Logic Quarterly*, 55(4):341–361, 2009.

[34] J. Wallenius, J. S. Dyer, P. C. Fishburn, R. E. Steuer, S. Zionts, and K. Deb. Multiple criteria decision making, multiattribute utility theory: Recent accomplishments and what lies ahead. *Management Science*, 54(7):1336–1349, 2008.

PREFERENCE INFERENCE BASED ON HIERARCHICAL AND SIMPLE LEXICOGRAPHIC MODELS

NIC WILSON, ANNE-MARIE GEORGE AND BARRY O'SULLIVAN
Insight Centre for Data Analytics, School of Computer Science and IT
University College Cork, Ireland
{nic.wilson, annemarie.george,
barry.osullivan}@insight-centre.org

Abstract

Preference Inference involves inferring additional user preferences from elicited or observed preferences, based on assumptions regarding the form of the user's preference relation. In this paper we consider a situation in which alternatives have an associated vector of costs, each component corresponding to a different criterion, and are compared using a kind of lexicographic order, similarly to the way alternatives are compared in a Hierarchical Constraint Logic Programming model. It is assumed that the user has some (unknown) importance ordering on criteria, and that to compare two alternatives, firstly, the combined cost of each alternative with respect to the most important criteria are compared; only if these combined costs are equal, are the next most important criteria considered. The preference inference problem then consists of determining whether a preference statement can be inferred from a set of input preferences. We show that this problem is coNP-complete, even if one restricts the cardinality of the equal-importance sets to have at most two elements, and one only considers non-strict preferences. However, it is polynomial if it is assumed that the user's ordering of criteria is a total ordering (which we call a simple lexicographic model); it is also polynomial if the sets of equally important criteria are all equivalence classes of a given fixed equivalence relation. We give an efficient polynomial algorithm for these cases, which also throws light on the structure of the inference. We give a complete proof theory for the simple lexicographic model case, and analyse variations of preference inference.[1]

1 Introduction

There are increasing opportunities for decision making/support systems to take into account the preferences of individual users, with the user preferences being elicited or observed from

---
[1]This is an extended version of an IJCAI-2015 paper [19].

the user's behaviour. However, users tend to have limited patience for preference elicitation, so such a system will tend to have a very incomplete picture of the user preferences. *Preference Inference* involves inferring additional user preferences from elicited or observed preferences, based on assumptions regarding the form of the user's preference relation. More specifically, given a set of input preferences Γ, and a set of preference models $\mathcal{M}$ (considered as candidates for the user's preference model), we infer a preference statement φ if every model in $\mathcal{M}$ that satisfies Γ also satisfies φ. Preference Inference can take many forms, depending on the choice of $\mathcal{M}$, and on the choices of language(s) for the input and inferred statements. For instance, if we just assume that the user model is a total order (or total pre-order), we can set $\mathcal{M}$ as the set of total [pre-]orders over a set of alternatives. This leads to a relatively cautious form of inference (based on transitive closure), including, for instance, the dominance relation for CP-nets and some related systems, e.g., [3, 5, 4, 1].

Often it can be valuable to obtain a much less cautious form of inference. In recommender systems for example, we aim to present the user with a relatively small set of alternatives. We can determine this set of alternatives as the undominated alternatives of a preference inference relation based on previously expressed user preferences [7, 14], with a more adventurous form of inference generating a smaller set of alternatives. Another example arises in a multi-objective context (as in a simple form of a Multi-Attribute Utility Theory model [9]). Again, it is often better if the number of optimal (undominated) solutions is relatively small, which can be achieved with a less cautious order relation on the set of objectives. These less cautious forms of inference include assuming that the user's preference relation is a simple weighted sum as considered in [7, 13, 12], or different lexicographic forms of preference models as in [16, 14, 18]. A comparison of Pareto orders, weighted sums and lexicographic orders in an multi-objective context shows that the lexicographic case is the least cautious and results in the least undominated solutions [12]. Note that all these systems involve reasoning about what holds in a set of preference models that coincide with the user's preference statements. This contrasts with work in preference learning that typically learns a single model, with the intention that this model closely resembles the real user's preference model [11, 8, 10, 6, 2].

In this paper we consider a situation in which alternatives have an associated vector of costs, each component corresponding to a different criterion, and are compared using a kind of lexicographic order, similarly to the way alternatives (feasible solutions) are compared in a Hierarchical Constraint Logic Programming (HCLP) model [15]. It is assumed that the user has some (unknown) importance ordering on criteria, and that to compare two alternatives, firstly, the combined cost of each alternative with respect to the most important criteria are compared; only if these combined costs are equal, are the next most important criteria considered. Implicitly, we assume that the costs of the alternatives are available to the user in order to express preference statements. Also, we assume to know all criteria the user might use and their costs.

We consider the case where the input preference statements are of a simple form that one alternative is preferred to another alternative, where we allow the expression of both strict and non-strict preferences (in contrast to most related preference logics, such as [17, 3, 16, 18] where only non-strict preferences are considered). We assume that the criteria by which the alternatives are compared are unfavorable facts like costs, distances, etc. Thus the lower values on the alternatives are the better. Accordingly, a strict preference $\alpha < \beta$ expresses that alternative α is better than β; a non-strict preference $\alpha \leq \beta$ means that α is at least as good as β. This form of preference is natural in many contexts, including for conversational recommender systems [7]. The preference inference problem then consists of determining whether a preference statement can be inferred from a set of input preferences, i.e., if every preference model (of the assumed form) satisfying the inputs also satisfies the query. We show that this problem is coNP-complete, even if one restricts the cardinality of the equal-importance sets to have at most two elements, and one only considers non-strict preferences. However, it is polynomial if it is assumed that the user's ordering of criteria is a total ordering (which we call the *simple lexicographic model* case); it is also polynomial if the sets of equally important criteria are all equivalence classes of a given fixed equivalence relation. We give an efficient polynomial algorithm for these cases, which also throws light on the structure of the inference.

Briefly, the idea behind the polynomial algorithm is as follows. Preference inference can be expressed in terms of testing consistency of a set of preference statements Γ. It turns out to be helpful to consider $\Gamma^{(\leq)}$, which is the same as Γ except that strict statements are replaced by non-strict ones on the same alternatives. We show that Γ is consistent if and only if some maximal model of $\Gamma^{(\leq)}$ satisfies Γ, which is if and only if every maximal model of $\Gamma^{(\leq)}$ satisfies Γ. Generating a maximal model of $\Gamma^{(\leq)}$ can be done in a simple and efficient way, using a greedy algorithm, thus allowing efficient testing of consistency (and thus preference inference). We also show that preference inference is compact, i.e., that if φ can be inferred from Γ then it can be inferred from a finite subset of Γ; and we analyse variations of preference inference, based on only considering maximal models, and only considering models that involve all the criteria.

We have defined our logics of preference inference in a semantic way. It is natural to consider whether we can define a complete proof theory, based on syntactic notion of consequence. We show how this can be done, if we extend the set of alternatives.

Section 2 defines our simple preference logic based on hierarchical models, along with some associated preference inference problems. Section 3 shows that in general the preference inference problem is coNP-complete. Section 4 considers the case where the importance ordering on criteria is a total order, and gives a polynomial algorithm for consistency; here we also consider variations of preference inference and relationships with a logic of disjunctive ordering constraints. In Section 5 we construct a complete proof theory, based on an extended set of alternatives. Section 6 concludes.

2 A preference logic based on hierarchical models

We consider preference models, based on an importance ordering of criteria, that is basically lexicographic, but involving a combination of criteria which are at the same level in the importance ordering. We call these "HCLP models", because models of a similar kind appear in the HCLP system [15] (though we have abstracted away some details from the latter system).

HCLP structures: Define an HCLP structure to be a tuple $\mathcal{S} = \langle \mathcal{A}, \oplus, \mathcal{C} \rangle$, where $\mathcal{A}$ (the set of *alternatives*) is a (possibly infinite) set; $\oplus$ is an associative, commutative and monotonic operation ($x \oplus y \leq z \oplus y$ if $x \leq z$) on the non-negative rational numbers $\mathbb{Q}^+$, with identity element 0; and $\mathcal{C}$ (known as the set of ($\mathcal{A}$-)*evaluations*) is a finite set[2] of functions from $\mathcal{A}$ to $\mathbb{Q}^+$. We also assume that operation $\oplus$ can be computed in linear time (which holds for natural definitions of $\oplus$, including addition and max). The evaluations in $\mathcal{C}$ may be considered as representing criteria or objectives under which the alternatives are evaluated. For $c \in \mathcal{C}$ and $\alpha \in \mathcal{A}$, if $c(\alpha) = 0$ then α fully satisfies the objective corresponding to c; more generally, the smaller the value of $c(\alpha)$, the better α satisfies the c-objective.

Example 1. *Suppose, a user wants to buy a new prepay mobile phone SIM card. She wants to make her decision between different providers based on the price per 10MB data usage d, the price per text message m and the price per minute for calls to the same provider c. These prices of d, m and c can be combined by addition. Consider four different options (providers) α, β, γ and δ with the following prices in cents.*

	α	β	γ	δ
d	18	15	13	14
m	15	17	15	13
c	10	11	14	15

In this context, the HCLP structure $\langle \mathcal{A}, \oplus, \mathcal{C} \rangle$ is given by the set of alternatives $\mathcal{A} = \{\alpha, \beta, \gamma, \delta\}$, the operator $\oplus$ being the ordinary addition on the integers and the set of evaluation functions $\mathcal{C} = \{d, m, c\}$.

---

[2] We could easily extend this to the case where $\mathcal{C}$ is a multi-set. (Or alternatively, we can reason about the latter case using the current formalism by adding an artificial alternative that every evaluation differs on.)

HCLP orderings: With each subset C of $\mathcal{C}$ we define ordering $\preccurlyeq_C^{\oplus}$ on $\mathcal{A}$ by $\alpha \preccurlyeq_C^{\oplus} \beta$ if and only if $\bigoplus_{c \in C} c(\alpha) \leq \bigoplus_{c \in C} c(\beta)$. Relation $\preccurlyeq_C^{\oplus}$ represents how well the alternatives satisfy the set of evaluations C if the latter are considered equally important. $\preccurlyeq_C^{\oplus}$ is a total pre-order (a weak order, i.e., a transitive and complete binary relation). We write $\equiv_C^{\oplus}$ for the associated equivalence relation on $\mathcal{A}$, given by $\alpha \equiv_C^{\oplus} \beta \iff \alpha \preccurlyeq_C^{\oplus} \beta$ and $\beta \preccurlyeq_C^{\oplus} \alpha$. We write $\prec_C^{\oplus}$ for the associated strict weak ordering, defined by $\alpha \prec_C^{\oplus} \beta \iff \alpha \preccurlyeq_C^{\oplus} \beta$ and $\beta \not\preccurlyeq_C^{\oplus} \alpha$. Thus, $\alpha \equiv_C^{\oplus} \beta$ if and only if $\bigoplus_{c \in C} c(\alpha) = \bigoplus_{c \in C} c(\beta)$; and $\alpha \prec_C^{\oplus} \beta$ if and only if $\bigoplus_{c \in C} c(\alpha) < \bigoplus_{c \in C} c(\beta)$.

HCLP models: An HCLP model H based on $\langle \mathcal{A}, \oplus, \mathcal{C} \rangle$ is defined to be an ordered partition $(C_1, \ldots, C_k)$ of a (possibly empty) subset of $\mathcal{C}$; we label this subset as $\sigma(H)$, so that $\sigma(H) = C_1 \cup \cdots \cup C_k$. The sets C_i are called the *levels of* H, which are thus non-empty, disjoint and have union $\sigma(H)$. If $c \in C_i$ and $c' \in C_j$, and $i < j$, then we say that c *appears before* c' (and c' *appears after* c) in H. Associated with H is an ordering relation $\preccurlyeq_H^{\oplus}$ on $\mathcal{A}$ given by:

$\alpha \preccurlyeq_H^{\oplus} \beta$ if and only if either:

(I) for all $i = 1, \ldots, k$, $\alpha \equiv_{C_i}^{\oplus} \beta$; or

(II) there exists some $i \in \{1, \ldots, k\}$ such that (i) $\alpha \prec_{C_i}^{\oplus} \beta$ and (ii) for all j with $1 \leq j < i$, $\alpha \equiv_{C_j}^{\oplus} \beta$.

Relation $\preccurlyeq_H^{\oplus}$ is a kind of lexicographic order on $\mathcal{A}$, where the set C_i of evaluations at the same level are first combined into a single evaluation. $\preccurlyeq_H^{\oplus}$ is a weak order on $\mathcal{A}$. We write $\equiv_H^{\oplus}$ for the associated equivalence relation (corresponding with condition (I)), and $\prec_H^{\oplus}$ for the associated strict weak order (corresponding with condition (II)), so that $\preccurlyeq_H^{\oplus}$ is the disjoint union of $\prec_H^{\oplus}$ and $\equiv_H^{\oplus}$. If $\sigma(H) = \emptyset$ then the first condition for $\alpha \preccurlyeq_H^{\oplus} \beta$ holds vacuously (since $k = 0$), so we have $\alpha \preccurlyeq_H^{\oplus} \beta$ for all $\alpha, \beta \in \mathcal{A}$, and $\prec_H^{\oplus}$ is the empty relation.

Preference language inputs: Let $\mathcal{A}$ be a set of alternatives. We define $\mathcal{L}_{\leq}^{\mathcal{A}}$ to be the set of statements of the form $\alpha \leq \beta$ ("α is preferred to β"), for $\alpha, \beta \in \mathcal{A}$ (the *non-strict* statements); we write $\mathcal{L}_{<}^{\mathcal{A}}$ for the set of statements of the form $\alpha < \beta$ ("α is strictly preferred to β"), for $\alpha, \beta \in \mathcal{A}$ (the *strict* statements); and we let $\mathcal{L}^{\mathcal{A}} = \mathcal{L}_{\leq}^{\mathcal{A}} \cup \mathcal{L}_{<}^{\mathcal{A}}$. If φ is the preference statement $\alpha \leq \beta$ then $\neg \varphi$ is defined to be the preference statement $\beta < \alpha$. If φ is the preference statement $\alpha < \beta$ then $\neg \varphi$ is defined to be the preference statement $\beta \leq \alpha$.

Satisfaction of preference statements: For an HCLP model H over the HCLP structure $\langle \mathcal{A}, \oplus, \mathcal{C} \rangle$, we say that H *satisfies* $\alpha \leq \beta$ (written $H \models^{\oplus} \alpha \leq \beta$) if $\alpha \preccurlyeq_H^{\oplus} \beta$ holds.

Similarly, we say that H *satisfies* $\alpha < \beta$ (written $H \models^\oplus \alpha < \beta$) if $\alpha \prec_H^\oplus \beta$. For $\Gamma \subseteq \mathcal{L}^{\mathcal{A}}$, we say that H *satisfies* Γ (written $H \models^\oplus \Gamma$) if H satisfies φ for all $\varphi \in \Gamma$. If $H \models^\oplus \varphi$ then we sometimes say that H *is a model of* φ (and similarly, if $H \models^\oplus \Gamma$).

Satisfaction of negated preference statements behaves as one would expect:

Lemma 1. *Let H be a HCLP model over HCLP structure $\mathcal{S}$. Then, H satisfies φ if and only if H does not satisfy $\neg\varphi$.*

Proof: Write $\mathcal{S}$ as $\langle \mathcal{A}, \oplus, \mathcal{C} \rangle$. It is sufficient to show that, for any $\alpha, \beta \in \mathcal{A}$, H satisfies $\alpha \leq \beta$ if and only if H does not satisfy $\beta < \alpha$. We have that H satisfies $\alpha \leq \beta$ if and only if $\alpha \preccurlyeq_H^\oplus \beta$, which, since $\preccurlyeq_H^\oplus$ is a weak order, is if and only if $\beta \not\prec_H^\oplus \alpha$, i.e., H does not satisfy $\beta < \alpha$. □

Example 2. *Consider Example 1 of a user choosing between different providers to buy a prepay SIM card. Suppose that the user is not interested in using data, and regards m and c as equally important. She can express her preferences by the corresponding HCLP model $H = (\{m, c\})$. Since $m(\alpha) + c(\alpha) = 25 < m(\beta) + c(\beta) = 28 = m(\delta) + c(\delta) = 28 < m(\gamma) + c(\gamma) = 29$, H satisfies $\alpha \prec_H^\oplus \beta \equiv_H^\oplus \delta \prec_H^\oplus \gamma$. The evaluations involved in H are $\sigma(H) = \{m, c\}$. If the user is most interested in the text message prices, and only if these are equal in the call prices, and only if these are also equal in the data prices, then the corresponding HCLP model is $H' = (\{m\}, \{c\}, \{d\})$. The induced order relation for this model satisfies $\delta \prec_{H'}^\oplus \alpha \prec_{H'}^\oplus \gamma \prec_{H'}^\oplus \beta$, since $m(\delta) < m(\alpha) = m(\gamma) < m(\beta)$ and $c(\alpha) < c(\gamma)$. The evaluations involved in H' are $\sigma(H') = \{d, m, c\}$.*

Preference inference/deduction relation: We are interested in different restrictions on the set of models, and the corresponding inference relations. Let $\mathcal{M}$ be a set of HCLP models over HCLP structure $\langle \mathcal{A}, \oplus, \mathcal{C} \rangle$. For $\Gamma \subseteq \mathcal{L}^{\mathcal{A}}$, and $\varphi \in \mathcal{L}^{\mathcal{A}}$, we say that $\Gamma \models_{\mathcal{M}}^\oplus \varphi$, if H satisfies φ for every $H \in \mathcal{M}$ satisfying Γ. Thus, if we elicit some preference statements Γ of a user, and we assume that their preference relation is an HCLP model in $\mathcal{M}$ (based on the HCLP structure), then $\Gamma \models_{\mathcal{M}}^\oplus \varphi$ holds if and only if we can deduce (with certainty) that the user's HCLP model H satisfies φ.

Consistency: For set of HCLP models $\mathcal{M}$ over HCLP structure $\langle \mathcal{A}, \oplus, \mathcal{C} \rangle$, and set of preference statements $\Gamma \subseteq \mathcal{L}^{\mathcal{A}}$, we say that Γ is $(\mathcal{M}, \oplus)$-*consistent* if there exists $H \in \mathcal{M}$ such that $H \models^\oplus \Gamma$; otherwise, we say that Γ is $(\mathcal{M}, \oplus)$-*inconsistent*. In the usual way, because of the existence of a negation operator, deduction can be reduced to checking (in)consistency.

Proposition 1. *$\Gamma \models_{\mathcal{M}}^\oplus \varphi$ if and only if $\Gamma \cup \{\neg\varphi\}$ is $(\mathcal{M}, \oplus)$-inconsistent.*

Proof: Suppose that $\Gamma \models^{\oplus}_{\mathcal{M}} \varphi$. By definition, H satisfies φ for every $H \in \mathcal{M}$ satisfying (every element of) Γ. Thus, using Lemma 1, there exists no $H \in \mathcal{M}$ that satisfies Γ and $\neg\varphi$, which implies that $\Gamma \cup \{\neg\varphi\}$ is $(\mathcal{M}, \oplus)$-inconsistent.

Conversely, suppose $\Gamma \cup \{\neg\varphi\}$ is $(\mathcal{M}, \oplus)$-inconsistent. By definition, there exists no $H \in \mathcal{M}$ that satisfies $\Gamma \cup \neg\varphi$. Thus, every $H \in \mathcal{M}$ that satisfies Γ does not satisfy $\neg\varphi$, and therefore satisfies φ, by Lemma 1. Hence, $\Gamma \models^{\oplus}_{\mathcal{M}} \varphi$. □

Let t be some number in $\{1, 2, \ldots, |\mathcal{C}|\}$. We define $\mathcal{C}(t)$ to be the set of all HCLP models $(C_1, \ldots, C_k)$ based on HCLP structure $\langle \mathcal{A}, \oplus, \mathcal{C} \rangle$ such that $|C_i| \leq t$, for all $i = 1, \ldots, k$. An element of $\mathcal{C}(1)$ thus corresponds to a sequence of singleton sets of evaluations; we identify it with a sequence of evaluations $(c_1, \ldots, c_k)$ in $\mathcal{C}$. Thus, $\Gamma \models^{\oplus}_{\mathcal{C}(t)} \varphi$ if and only if $H \models^{\oplus} \varphi$ for all $H \in \mathcal{C}(t)$ such that $H \models^{\oplus} \Gamma$. Note that for $t = 1$, these definitions do not depend on $\oplus$ (since there is no combination of evaluations involved), so we may drop any mention of $\oplus$.

Let $\equiv$ be an equivalence relation on $\mathcal{C}$, and let $\mathcal{E}$ be the set of equivalence classes of $\equiv$. Thus, for each $c \in \mathcal{C}$ there exists a unique element $E \in \mathcal{E}$ such that $E \ni c$, and $E = \{c' \in \mathcal{C} : c' \equiv c\}$. We define $\mathcal{C}(\equiv)$ to be the set of all HCLP models $(C_1, \ldots, C_k)$ such that each C_i is an equivalence class with respect to $\equiv$, i.e., $C_i \in \mathcal{E}$. It is easy to see that the relation $\models^{\oplus}_{\mathcal{C}(\equiv)}$ is the same as the relation $\models_{C'(1)}$ where C' is defined as follows. C' is in 1-1 correspondence with $\mathcal{E}$. If E is the $\equiv$-equivalence class of $\mathcal{C}$ corresponding with $c' \in C'$ then, for $\alpha \in \mathcal{A}$, $c'(\alpha)$ is defined to be $\bigoplus_{c \in E} c(\alpha)$, so that each C_i in an HCLP model is replaced by a single evaluation equivalent to the combination of all the elements of C_i.

For $\models$ either being $\models^{\oplus}_{\mathcal{C}(t)}$ for some $t \in \{1, 2, \ldots, |\mathcal{C}|\}$, or being $\models^{\oplus}_{\mathcal{C}(\equiv)}$ for some equivalence relation $\equiv$ on $\mathcal{C}$, we consider the following decision problem.

HCLP-DEDUCTION FOR $\models$: Given $\mathcal{C}$, Γ and φ is it the case that $\Gamma \models \varphi$?

In Section 4, we will show that this problem is polynomial for $\models$ being $\models^{\oplus}_{\mathcal{C}(t)}$ when $t = 1$. Thus it is polynomial also for $\models^{\oplus}_{\mathcal{C}(\equiv)}$, for any equivalence relation $\equiv$. It is coNP-complete for $\models$ being $\models^{\oplus}_{\mathcal{C}(t)}$ when $t > 1$, as shown below in Section 3.

Theorem 1. *HCLP-DEDUCTION FOR $\models^{\oplus}_{\mathcal{C}(t)}$ is polynomial when $t = 1$, and is coNP-complete for any $t > 1$, even if we restrict the language to non-strict preference statements. HCLP-DEDUCTION FOR $\models^{\oplus}_{\mathcal{C}(\equiv)}$ is polynomial for any equivalence relation $\equiv$.*

Example 3. *Consider the HCLP structure of Example 1. Suppose, the user states that she prefers α to β, i.e. $\alpha \leq \beta$, and strictly prefers β to γ, i.e. $\beta < \gamma$. Only the HCLP models of the forms $(\{c\}, \ldots)$, $(\{m\}, \ldots)$, $(\{c, m\}, \ldots)$ or $(\{d, m, c\})$ satisfy $\alpha \leq \beta$. Only the HCLP models $(\{c\}, \ldots)$, $(\{c, d\}, \ldots)$ or $(\{c, m\}, \ldots)$ satisfy $\beta < \gamma$. Thus, the models $(\{c\}, \ldots)$ and $(\{c, m\}, \ldots)$ are the only ones that satisfy the set $\Gamma = \{\alpha \leq \beta, \beta < \gamma\}$*

of the user's input preferences. Let $t \in \{1,2,3\}$. Then $\Gamma \not\models^{\oplus}_{\mathcal{C}(t)} \delta \leq \beta$ since the model $H = (\{c\}) \in \mathcal{C}(1) \subseteq \mathcal{C}(t)$ satisfies Γ and $\beta \prec^{\oplus}_{H} \delta$, i.e., $H \not\models^{\oplus} \delta \leq \beta$. Furthermore, $\Gamma \not\models^{\oplus}_{\mathcal{C}(2)} \beta \leq \delta$ since the model $H' = (\{c,m\},\{d\}) \in \mathcal{C}(2)$ satisfies Γ and $\delta \prec^{\oplus}_{H'} \beta$, i.e., $H' \not\models^{\oplus} \beta \leq \delta$. However, we can infer $\Gamma \models^{\oplus}_{\mathcal{C}(1)} \beta \leq \delta$, and even $\Gamma \models^{\oplus}_{\mathcal{C}(1)} \beta < \delta$, since all Γ-satisfying HCLP models in $\mathcal{C}(1)$, i.e., $(\{c\})$, $(\{c\},\{m\})$, $(\{c\},\{d\})$, $(\{c\},\{m\},\{d\})$, and $(\{c\},\{d\},\{m\})$, satisfy the relation $\beta < \delta$.

3 Proving coNP-completeness of HCLP-deduction for $\models^{\oplus}_{\mathcal{C}(t)}$ for $t > 1$

Given an arbitrary 3-SAT instance we will show that we can construct a set Γ and a statement $\alpha \leq \beta$ such that the 3-SAT instance has a satisfying truth assignment if and only if $\Gamma \not\models^{\oplus}_{\mathcal{C}(t)} \alpha \leq \beta$ (see Proposition 2 below). This then implies that determining if $\Gamma \not\models^{\oplus}_{\mathcal{C}(t)} \alpha \leq \beta$ holds is NP-hard.

We have that $\Gamma \not\models^{\oplus}_{\mathcal{C}(t)} \alpha \leq \beta$ if and only if there exists an HCLP-model $H \in \mathcal{C}(t)$ such that $H \models^{\oplus} \Gamma$ and $H \not\models^{\oplus} \alpha \leq \beta$. For any given H, checking that $H \models^{\oplus} \Gamma$ and $H \not\models^{\oplus} \alpha \leq \beta$ can be performed in polynomial time. This implies that determining if $\Gamma \not\models^{\oplus}_{\mathcal{C}(t)} \alpha \leq \beta$ holds is in NP, and therefore is NP-complete, and thus determining if $\Gamma \models^{\oplus}_{\mathcal{C}(t)} \alpha \leq \beta$ holds is coNP-complete.

Consider an arbitrary 3-SAT instance based on propositional variables $p_1, \ldots, p_r$, consisting of clauses Λ_j, for $j = 1, \ldots, s$. For each propositional variable p_i we associate two evaluations q_i^+ and q_i^-, where q_i^- corresponds with literal $\neg p_i$, and q_i^+ corresponds with literal p_i.

The idea behind the construction is as follows: we generate a (polynomial size) set $\Gamma \subseteq \mathcal{L}^{\mathcal{A}}_{\leq}$ as the disjoint union of sets Γ_1, Γ_2 and Γ_3, and we choose a non-strict statement $\alpha \leq \beta$. For the remainder of this section, let H be an arbitrary HCLP-model in $\mathcal{C}(t)$. Γ_1 is chosen so that if $H \models^{\oplus} \Gamma_1$ then, for each $i = 1, \ldots, r$, $\sigma(H)$ cannot contain both q_i^+ and q_i^-, i.e., q_i^+ and q_i^- do not both appear in H. (Recall H is an ordered partition of $\sigma(H)$, so that $\sigma(H)$ is the subset of $\mathcal{C}$ that appears in H.) If $H \models^{\oplus} \Gamma_2$ and $H \models^{\oplus} \beta < \alpha$ then $\sigma(H)$ contains either q_i^+ or q_i^-. Together, this implies that if $H \models^{\oplus} \Gamma$ and $H \not\models^{\oplus} \alpha \leq \beta$ then for each propositional variable p_i, model H involves either q_i^+ or q_i^-, but not both. Γ_3 is used to make the correspondence with the clauses. For instance, if one of the clauses is $p_2 \vee \neg p_5 \vee p_6$ then any HCLP model $H \in \mathcal{C}(t)$ of $\Gamma \cup \{\beta < \alpha\}$ will involve either q_2^+, q_5^-, or q_6^+.

Suppose that H satisfies Γ but not $\alpha \leq \beta$. We can generate a satisfying assignment of the 3-SAT instance, by assigning p_i to 1 (TRUE) if and only if q_i^+ appears in H.

The monotonicity assumption for operation $\oplus$ implies that $1 \oplus 1 > 0$, since we have $1 \oplus 1 \geq 1 \oplus 0 = 1 > 0$. In fact, in the proof below we do not need to assume monotonicity of $\oplus$; it is sufficient to just assume that $1 \oplus 1 > 0$.

We describe the construction more formally below.

Defining $\mathcal{A}$ and $\mathcal{C}$: The set of alternatives $\mathcal{A}$ is defined to be the union of the following sets

- $\{\alpha, \beta\} \cup \{\alpha_i, \beta_i, \delta_i : i = 1, \ldots, r\}$
- $\{\gamma_i^k : i = 1, \ldots, r, k = 1, \ldots, t-1\}$
- $\{\theta_j, \tau_j : j = 1, \ldots, s\}$.

We define the set of evaluations $\mathcal{C}$ to be $\{c^*\} \cup \{q_i^+, q_i^- : i = 1, \ldots, r\} \cup A_1 \cup \cdots \cup A_r$, where $A_i = \{a_i^k : k = 1, \ldots, t-1\}$. Both $\mathcal{A}$ and $\mathcal{C}$ are of polynomial size.

Satisfying $\beta < \alpha$: The evaluations on α and β are defined as follows:

- $c^*(\alpha) = 1$, and for all $c \in \mathcal{C} - \{c^*\}$, $c(\alpha) = 0$.
- For all $c \in \mathcal{C}$, $c(\beta) = 0$.

It immediately follows that: $H \models^\oplus \beta < \alpha \iff \sigma(H) \ni c^*$.

The construction of Γ_1: We define $\Gamma_1 = \bigcup_{i=1}^r \Gamma_1^i$ where, for each $i = 1, \ldots, r$, we define $\Gamma_1^i = \{\delta_i \leq \gamma_i^k, \gamma_i^k \leq \delta_i : k = 1, \ldots, t-1\}$. We make use of auxiliary evaluations $A_i = \{a_i^1, \ldots, a_i^{t-1}\}$. The values of the evaluations on γ_i^k and δ_i are defined as follows:

- $a_i^k(\gamma_i^k) = 1$, and for all $c \in \mathcal{C} - \{a_i^k\}$ we set $c(\gamma_i^k) = 0$.
- $q_i^+(\delta_i) = q_i^-(\delta_i) = 1$, and for other $c \in \mathcal{C}$, $c(\delta_i) = 0$.

Thus, for any $B \subseteq A_i$, we have $(\bigoplus_{a \in B} a \oplus q_i^+)(\delta_i) = \bigoplus_{a \in B} a(\delta_i) \oplus q_i^+(\delta_i) = 0 \oplus \cdots \oplus 0 \oplus 1 = 1$. Similarly, $(\bigoplus_{a \in B} a \oplus q_i^-)(\delta_i) = 1$. Furthermore, $(\bigoplus_{a \in B} a \oplus q_i^+)(\gamma_i^k) = 1 \iff a_i^k \in B$ and $(\bigoplus_{a \in B} a \oplus q_i^-)(\gamma_i^k) = 1 \iff a_i^k \in B$.

Lemma 2. $H \models^\oplus \Gamma_1^i$ if and only if either (i) $\sigma(H)$ does not contain any element in A_i or q_i^+ or q_i^-, i.e., $\sigma(H) \cap (A_i \cup \{q_i^+, q_i^-\}) = \emptyset$; or (ii) $A_i \cup \{q_i^+\}$ is a level of H, and $\sigma(H) \not\ni q_i^-$; or (iii) $A_i \cup \{q_i^-\}$ is a level of H, and $\sigma(H) \not\ni q_i^+$. In particular, if $H \models^\oplus \Gamma_1^i$ then $\sigma(H)$ does not contain both q_i^+ and q_i^-.

Proof: Consider any $H \in \mathcal{C}(t)$, so that for each level E of H we have $|E| \le t$. We have that $H \models^\oplus \Gamma_1^i$ if and only if for each level E of H and for all $k = 1, \ldots, t-1$, $\delta_i \equiv_E^\oplus \gamma_i^k$. Now, $\delta_i \equiv_E^\oplus \gamma_i^k$ if and only if $\bigoplus_{c \in E} c(\delta_i) = \bigoplus_{c \in E} c(\gamma_i^k)$. Also, $\bigoplus_{c \in E} c(\delta_i) = 0$ unless E contains either q_i^+ or q_i^-; and $\bigoplus_{c \in E} c(\delta_i) = 1 \oplus 1 > 0$ if E contains both q_i^+ and q_i^-; and equals 1 if E contains either q_i^+ or q_i^-, but not both. $\bigoplus_{c \in E} c(\gamma_i^k)$ equals 1 if and only if E contains a_i^k, and equals 0 otherwise.

This implies that if for all $k = 1, \ldots, t-1$, $\delta_i \equiv_E^\oplus \gamma_i^k$ and E contains q_i^+ or q_i^- then for all $k = 1, \ldots, t-1$, E contains a_i^k, and so $E \supseteq A_i$. Because of the condition that $|E| \le t$ (since $H \in \mathcal{C}(t)$), and $|A_i| = t - 1$, we then have that E equals either $A_i \cup \{q_i^+\}$ or $A_i \cup \{q_i^-\}$.

Similarly, if for all $k = 1, \ldots, t-1$, $\delta_i \equiv_E^\oplus \gamma_i^k$ and E contains a_i^k for some $k \in \{1, \ldots, t-1\}$, then E contains q_i^+ or q_i^-, and so, by the previous paragraph, E equals either $A_i \cup \{q_i^+\}$ or $A_i \cup \{q_i^-\}$.

Thus, if $H \models^\oplus \Gamma_1^i$, then for at most one level E of H do we have $E \cap (A_i \cup \{q_i^+, q_i^-\})$ non-empty (else we would have two levels both containing A_i, contradicting disjointness of levels); also if $E \cap (A_i \cup \{q_i^+, q_i^-\})$ is non-empty then E equals either $A_i \cup \{q_i^+\}$ or $A_i \cup \{q_i^-\}$. In particular, if $H \models^\oplus \Gamma_1^i$ then $\sigma(H)$ does not contain both q_i^+ and q_i^-.

Regarding the converse, let us suppose first that (i) $\sigma(H)$ does not intersect with $A_i \cup \{q_i^+, q_i^-\}$. Then for all levels E of H, and for all $k = 1, \ldots, t-1$, we have $\bigoplus_{c \in E} c(\delta_i) = \bigoplus_{c \in E} c(\gamma_i^k) = 0$, and thus $\delta_i \equiv_E^\oplus \gamma_i^k$, which implies $H \models^\oplus \Gamma_1^i$.

Now suppose (ii) that $A_i \cup \{q_i^+\}$ is a level E' of H and $\sigma(H) \not\ni q_i^-$. Then every other level E is disjoint from $A_i \cup \{q_i^+, q_i^-\}$, so for all $k = 1, \ldots, t-1$, $\bigoplus_{c \in E} c(\delta_i) = \bigoplus_{c \in E} c(\gamma_i^k) = 0$, and thus $\delta_i \equiv_E^\oplus \gamma_i^k$. Also, $\bigoplus_{c \in E'} c(\delta_i) = \bigoplus_{c \in E'} c(\gamma_i^k) = 1$, and thus $H \models^\oplus \Gamma_1^i$. Case (iii), when $A_i \cup \{q_i^-\}$ is a level E' of H and $\sigma(H) \not\ni q_i^+$, is essentially identical to Case (ii), just switching the roles of q_i^+ and q_i^+. $\square$

The construction of Γ_2: For each $i = 1, \ldots, r$, define φ_i to be $\alpha_i \le \beta_i$. We let $\Gamma_2 = \{\varphi_i : i = 1, \ldots, r\}$. The values of the evaluations on α_i and β_i are defined as follows. We define $c^*(\alpha_i) = 1$, and for all $c \in \mathcal{C} - \{c^*\}$, $c(\alpha_i) = 0$. Define $q_i^+(\beta_i) = q_i^-(\beta_i) = 1$, and for all $c \in \mathcal{C} - \{q_i^+, q_i^-\}$, $c(\beta_i) = 0$. Thus, similarly to the previous observations for Γ_1, $(c^* \oplus q_i^+)(\beta_i) = (c^* \oplus q_i^-)(\beta_i) = 1$ and $(c^* \oplus q_i^+)(\alpha_i) = (c^* \oplus q_i^-)(\alpha_i) = 1$. Also, $(q_i^+ \oplus q_i^-)(\alpha_i) = 0$ and $(q_i^+ \oplus q_i^-)(\beta_i) \ge 1$, because of the monotonicity of $\oplus$, and $(c^* \oplus q_i^+ \oplus q_i^-)(\alpha_i) = 1$ and $(c^* \oplus q_i^+ \oplus q_i^-)(\beta_i) \ge 1$.

The following result easily follows.

Lemma 3. *If q_i^+ or q_i^- appears before c^* in H then $H \models^\oplus \varphi_i$. If $\sigma(H) \ni c^*$ and $H \models^\oplus \varphi_i$ then $\sigma(H) \ni q_i^+$ or $\sigma(H) \ni q_i^-$.*

Proof: Consider any $H \in \mathcal{C}(t)$, and consider any $i \in \{1, \ldots, r\}$. Then the following hold for any level E of H.

(I) If E does not contain any of $\{c^*, q_i^+, q_i^-\}$ then $\bigoplus_{c \in E} c(\alpha_i) = \bigoplus_{c \in E} c(\beta_i) = 0$ so $\alpha_i \equiv_E^\oplus \beta_i$.

(II) If E contains c^* but neither of q_i^+ or q_i^-, then $\bigoplus_{c \in E} c(\alpha_i) = 1$ and $\bigoplus_{c \in E} c(\beta_i) = 0$, so $\alpha_i \not\equiv_E^\oplus \beta_i$.

(III) If E contains q_i^+ or q_i^- but not c^* then $\bigoplus_{c \in E} c(\alpha_i) = 0$ and $\bigoplus_{c \in E} c(\beta_i) > 0$ using the fact that $1 \oplus 1 > 0$, so $\alpha_i \prec_E^\oplus \beta_i$.

Assume that $\sigma(H) \ni c^*$. If $\sigma(H) \cap \{q_i^+, q_i^-\} = \emptyset$ then by considering the level containing c^* we can see, using (I) and (II), that $\alpha_i \not\equiv_H^\oplus \beta_i$, so $H \not\models^\oplus \varphi_i$. This proves the second half of the lemma.

If q_i^+ or q_i^- (or both) appear before c^* in H then (I) and (III) imply that $\alpha_i \prec_H^\oplus \beta_i$ and thus $H \models^\oplus \varphi_i$. □

The construction of Γ_3: For each $i = 1, \ldots, r$, define $Q(p_i) = q_i^+$ and $Q(\neg p_i) = q_i^-$. This defines the function Q over all literals. Let us write the jth clause as $l_1 \vee l_2 \vee l_3$ for literals l_1, l_2 and l_3. Define $Q_j = \{Q(l_1), Q(l_2), Q(l_3)\}$. For example, if the jth clause were $p_2 \vee \neg p_5 \vee p_6$ then $Q_j = \{q_2^+, q_5^-, q_6^+\}$. We define ψ_j to be $\theta_j \le \tau_j$, and $\Gamma_3 = \{\psi_j : j = 1, \ldots s\}$. Define $c^*(\theta_j) = 1$ and $c(\theta_j) = 0$ for all $c \in \mathcal{C} - \{c^*\}$. Define $q(\tau_j) = 1$ for $q \in Q_j$, and for all other c (i.e., $c \in \mathcal{C} - Q_j$), define $c(\tau_i) = 0$.

Lemma 4. *If some element of Q_j appears in H before c^*, and no level of H contains more than one element of Q_j, then $H \models^\oplus \psi_j$. If $\sigma(H) \ni c^*$ and $H \models^\oplus \psi_j$ then $\sigma(H)$ contains some element of Q_j.*

Proof: The proof of this result is similar to that of Lemma 3. Consider any $H \in \mathcal{C}(t)$ any clause j. Then the following hold for any level E of H.

(I) If E does not contain any element of $Q_j \cup \{c^*\}$ then $\bigoplus_{c \in E} c(\theta_j) = \bigoplus_{c \in E} c(\tau_j) = 0$ so $\theta_j \equiv_E^\oplus \tau_j$.

(II) If E contains c^* but no element of Q_j neither of q_i^+ or q_i^-, then $\bigoplus_{c \in E} c(\theta_j) = 1$ and $\bigoplus_{c \in E} c(\tau_j) = 0$, so $\theta_j \not\equiv_E^\oplus \tau_j$.

(III) If E contains exactly one element of Q_j but not c^* then $\bigoplus_{c \in E} c(\theta_j) = 0$ and $\bigoplus_{c \in E} c(\tau_j) = 1$, so $\theta_j \prec_E^\oplus \tau_j$.

Assume that $\sigma(H) \ni c^*$. If $\sigma(H) \cap Q_j = \emptyset$ then by considering the level containing c^* we can see, using (I) and (II), that $\theta_j \not\prec_H^\oplus \tau_j$, so $H \not\models^\oplus \varphi_i$. This argument proves that if $\sigma(H) \ni c^*$ and $H \models^\oplus \psi_j$ then $\sigma(H)$ contains some element of Q_j.

If some element of Q_j appears in H before c^*, and no level of H contains more than one element of Q_j, then (I) and (III) imply that $\theta_j \prec_H^\oplus \tau_j$ and thus $H \models^\oplus \varphi_i$. □

We set $\Gamma = \Gamma_1 \cup \Gamma_2 \cup \Gamma_3$. The following result implies that the HCLP deduction problem is coNP-hard (even if we restrict to the case when $\Gamma \cup \{\varphi\} \subseteq \mathcal{L}_\leq^\mathcal{A}$).

Proposition 2. *Using the notation defined above, the 3-SAT instance is satisfiable if and only if $\Gamma \not\models_{\mathcal{C}(t)}^\oplus \alpha \leq \beta$.*

Proof: First let us assume that $\Gamma \not\models_{\mathcal{C}(t)}^\oplus \alpha \leq \beta$. Then by definition, there exists an HCLP model $H \in \mathcal{C}(t)$ with $H \models^\oplus \Gamma$ and $H \not\models^\oplus \alpha \leq \beta$. Since $H \not\models^\oplus \alpha \leq \beta \iff H \models^\oplus \beta < \alpha$, we have $H \models^\oplus \Gamma \cup \{\beta < \alpha\}$. Because $H \models^\oplus \beta < \alpha$, we have $\sigma(H) \ni c^*$.

Because also $H \models^\oplus \Gamma_2^i$, either $\sigma(H) \ni q_i^+$ or $\sigma(H) \ni q_i^-$, by Lemma 3. Since $H \models^\oplus \Gamma_1^i$, the set $\sigma(H)$ does not contain both q_i^+ and q_i^-, by Lemma 2.

Let us define a truth function $f : \mathcal{P} \to \{0, 1\}$ as follows: $f(p_i) = 1 \iff \sigma(H) \ni q_i^+$. Since $\sigma(H)$ contains exactly one of q_i^+ and q_i^-, we have $f(p_i) = 0 \iff \sigma(H) \ni q_i^-$. We extend f to negative literals in the obvious way: $f(\neg p_i) = 1 - f(p_i)$, and thus, $f(\neg p_i) = 1 \iff \sigma(H) \ni q_i^-$.

Since $H \models^\oplus \Gamma_3$ and $\sigma(H) \ni c^*$, then $\sigma(H)$ contains at least one element of each Q_j, by Lemma 4. Thus for each j, $f(l) = 1$ for at least one literal l in the jth clause, and hence f satisfies clause Λ_j. We have shown that f satisfies each clause of the 3-SAT instance, proving that the instance is satisfiable.

Conversely, suppose that the 3-SAT instance is satisfiable, so there exists a truth function f satisfying it. We will construct an HCLP model $H \in \mathcal{C}(t)$ such that $H \models^\oplus \Gamma \cup \{\beta < \alpha\}$, and thus $H \not\models^\oplus \alpha \leq \beta$, proving that $\Gamma \not\models_{\mathcal{C}(t)}^\oplus \alpha \leq \beta$.

For $i = 1, \ldots, r$, let $S_i = A_i \cup \{q_i^+\}$ if $f(p_i) = 1$, and otherwise, let $S_i = A_i \cup \{q_i^-\}$. Thus, if $f(p_i) = 1$ then $Q(p_i) \in S_i$; and if $f(\neg p_i) = 1$ then $Q(\neg p_i) \in S_i$. We then define H to be the sequence $S_1, S_2, \ldots, S_r, \{c^*\}$. Since $\sigma(H) \ni c^*$, we have that $H \models^\oplus \beta < \alpha$. By Lemma 2, for all $i = 1, \ldots, r$, $H \models^\oplus \Gamma_1^i$ and so $H \models^\oplus \Gamma_1$. By Lemma 3, for all $i = 1, \ldots, r$, $H \models^\oplus \varphi_i$, so $H \models^\oplus \Gamma_2$.

Consider any $j \in \{1, \ldots, s\}$, and, as above, write the jth clause as $l_1 \vee l_2 \vee l_3$. Truth assignment f satisfies this clause, so there exists $k \in \{1, 2, 3\}$ such that $f(l_k) = 1$. Then $Q(l_k)$ appears in H before c^*, so, by Lemma 4, $H \models^\oplus \psi_j$. Thus $H \models^\oplus \Gamma_3$. Since $\Gamma = \Gamma_1 \cup \Gamma_2 \cup \Gamma_3$, we have shown that $H \models^\oplus \Gamma \cup \{\beta < \alpha\}$, proving that $\Gamma \not\models_{\mathcal{C}(t)}^\oplus \alpha \leq \beta$. □

Example 4. Let $(p_1 \vee p_2 \vee \neg p_3) \wedge (\neg p_1 \vee p_2 \vee p_3)$ be an instance of 3-SAT with the three propositional variables p_1, p_2, p_3 and clauses Λ_1, Λ_2. From this we construct a $\mathcal{C}(2)$ HCLP-Deduction instance as in the previous paragraphs (so with $t = 2$). Corresponding to the two possible assignments of each of the propositional variables p_1, p_2, p_3, we construct evaluation functions q_1^+, q_2^+, q_3^+ and q_1^-, q_2^-, q_3^-. We also introduce the additional evaluation functions c^* and $A_1 = \{a_1^1, a_2^1, a_3^1\}$. Furthermore, we construct alternatives $\alpha, \beta, \alpha_1, \alpha_2, \alpha_3, \beta_1, \beta_2, \beta_3, \delta_1, \delta_2, \delta_3, \gamma_1^1, \gamma_2^1, \gamma_3^1, \theta_1, \theta_2, \tau_1, \tau_2$ for the preference statements $\alpha > \beta, \Gamma_1, \Gamma_2$ and Γ_3, with the values of the evaluation functions given as follows:

	$\alpha > \beta$	$\alpha_1 \leq \beta_1$		$\alpha_2 \leq \beta_2$		$\alpha_3 \leq \beta_3$		$\delta_1 \leq,\geq \gamma_1^1$		$\delta_2 \leq,\geq \gamma_2^1$		$\delta_3 \leq,\geq \gamma_3^1$		$\theta_1 \leq \tau_1$		$\theta_2 \leq \tau_2$	
								Γ_2			Γ_1					Γ_3	
q_1^+	0	0	0	1	0	0	0	0	1	0	0	0	0	0	0	1	0	0
q_2^+	0	0	0	0	0	1	0	0	0	0	1	0	0	0	0	1	0	1
q_3^+	0	0	0	0	0	0	0	1	0	0	0	0	1	0	0	0	0	1
q_1^-	0	0	0	1	0	0	0	0	1	0	0	0	0	0	0	0	0	1
q_2^-	0	0	0	0	0	1	0	0	0	0	1	0	0	0	0	0	0	0
q_3^-	0	0	0	0	0	0	0	1	0	0	0	0	1	0	0	1	0	0
c^*	1	0	1	0	1	0	1	0	0	0	0	0	0	0	1	0	1	0
a_1^1	0	0	0	0	0	0	0	0	0	1	0	0	0	0	0	0	0	0
a_2^1	0	0	0	0	0	0	0	0	0	0	0	1	0	0	0	0	0	0
a_3^1	0	0	0	0	0	0	0	0	0	0	0	0	0	1	0	0	0	0

Here, the values of τ_1 and τ_2 correspond to the occurrences of the literals p_i or $\neg p_i$ in the clauses Λ_1 and Λ_2, respectively. Since the statement $\alpha > \beta$ is strict, the evaluation c^* has to be included in any satisfying HCLP model H of $\Gamma \cup \{\alpha > \beta\}$, where $\Gamma = \Gamma_1 \cup \Gamma_2 \cup \Gamma_3$. To satisfy a non-strict preference statement $\nu \leq \rho$, if a level contains an evaluation function with value 1 for ν then the same or an earlier level must contain an evaluation function with value 1 for ρ. The preference statement e.g., $\alpha_1 \leq \beta_1$ in Γ_2 then enforces that either q_1^+ or q_1^- appears in some level of H (and no later than c^*) because $c^*(\alpha_1) = 1$ and $q_1^+(\beta_1) = q_1^-(\beta_1) = 1$. Since Γ_1 contains $\delta_1 \leq \gamma_1^1$ and $\gamma_1^1 \leq \delta_1$, a $\mathcal{C}(2)$-HCLP model H satisfying $\Gamma \cup \{\alpha > \beta\}$ must have a_1^1 appearing in the same level as q_1^+ or q_1^-, and both q_1^+ and q_1^- cannot then appear in H. Thus H involves either q_1^+ or q_1^- but not both. Γ_3 contains ψ_1, i.e., $\theta_1 \leq \tau_1$, which ensures that at least one element in $Q_1 = \{q_1^+, q_2^+, q_3^-\}$ appears in some level of a satisfying HCLP model, which corresponds to satisfying the first clause. The assignment $p_1 = $ true, $p_2 = $ true, $p_3 = $ false satisfies the instance $(p_1 \vee p_2 \vee \neg p_3) \wedge (\neg p_1 \vee p_2 \vee p_3)$. A corresponding $\Gamma \cup \{\alpha > \beta\}$-satisfying HCLP model in $\mathcal{C}(2)$ is $(\{q_1^+, a_1^1\}, \{q_2^+, a_2^1\}, \{q_3^-, a_3^1\}, \{c^*\})$.

4 Simple lexicographic models

In this section, we consider the case where we restrict to HCLP models which consist of a sequence of singletons; thus each model corresponds to a sequence of evaluations, and generates a lexicographic order based on these. We call such models: *simple lexicographic models*.

Let $\mathcal{C}$ be a set of evaluations on $\mathcal{A}$. To simplify notation, we redefine a $\mathcal{C}(1)$-model to be a sequence of different elements of $\mathcal{C}$ (rather than a sequence of singleton sets). As mentioned earlier, the operation $\oplus$ plays no part, so we can harmlessly abbreviate ordering $\preccurlyeq_H^\oplus$ to just $\preccurlyeq_H$, for any $\mathcal{C}(1)$-model H, and similarly for $\prec_H$ and $\equiv_H$. The deduction problem for the sequence of singletons case is thus as follows. Given $\Gamma \cup \{\varphi\} \subseteq \mathcal{L}^\mathcal{A}$, is it the case that $\Gamma \models_{\mathcal{C}(1)} \varphi$? That is, is it the case that for all $\mathcal{C}(1)$-models H (over $\mathcal{A}$), if H satisfies Γ then H satisfies φ?

Given set of evaluations $\mathcal{C}$ and set of preference statements Γ, we introduce in Section 4.1 the important concept of *maximal inconsistency base* $(\Gamma^\perp, C^\perp)$, where $\Gamma^\perp \subseteq \Gamma$ and $C^\perp \subseteq \mathcal{C}$. No model of Γ involves any element of $C^\perp$, and it turns out (Corollary 1) that Γ is $\mathcal{C}(1)$-inconsistent if and only if $\Gamma^\perp$ contains a strict element. It is helpful (see Section 4.2) to consider $\Gamma^{(\leq)}$, a version of Γ where each strict element is replaced by the corresponding non-strict one. Models of $\Gamma^{(\leq)}$ can be generated in a simple iterative way. If one model of $\Gamma^{(\leq)}$ extends another, then the former satisfies at least as many elements of Γ as the latter does. It is natural to then consider maximal models of $\Gamma^{(\leq)}$. We show (Proposition 8) that maximal models of $\Gamma^{(\leq)}$ involve every evaluation except the ones in $C^\perp$, and satisfy every element of Γ except the strict statements in $\Gamma^\perp$. This implies that all maximal models of $\Gamma^{(\leq)}$ involve the same evaluations and satisfy the same subset of Γ. Thus to determine if Γ is $\mathcal{C}(1)$-consistent, we just have to generate any maximal model of $\Gamma^{(\leq)}$ (see Theorems 2 and 3), which can be done with a simple greedy algorithm, and test if this model satisfies Γ.

A nice mathematical property of this form of preference inference is compactness (see Corollary 2): any inference from an infinite set Γ also follows from some finite subset of it.

Our notion of preference inference is an intuitive one; however, there are also natural variations based on only considering models that involve all the evaluations; or alternatively, only considering maximal models. We explore such variations of preference inference in Section 4.3, and show strong connections with the main notion of preference inference. In Section 4.4 we show how the preference inference based on simple lexicographic models is very closely related to a logic based on disjunctive ordering statements.

4.1 Some basic definitions and results

We write $\varphi \in \mathcal{L}^{\mathcal{A}}$ as $\alpha_\varphi < \beta_\varphi$, if φ is strict, or as $\alpha_\varphi \leq \beta_\varphi$, if φ is non-strict. We consider a set $\Gamma \subseteq \mathcal{L}^{\mathcal{A}}$, and a set $\mathcal{C}$ of evaluations on $\mathcal{A}$.

Supp$_\mathcal{C}^\varphi$, Opp$_\mathcal{C}^\varphi$ and Ind$_\mathcal{C}^\varphi$: For $\varphi \in \Gamma$, define *Supp$_\mathcal{C}^\varphi$* to be $\{c \in \mathcal{C} : c(\alpha_\varphi) < c(\beta_\varphi)\}$; define *Opp$_\mathcal{C}^\varphi$* to be $\{c \in \mathcal{C} : c(\alpha_\varphi) > c(\beta_\varphi)\}$; and define *Ind$_\mathcal{C}^\varphi$* to be $\{c \in \mathcal{C} : c(\alpha_\varphi) = c(\beta_\varphi)\}$. Thus, *Supp$_\mathcal{C}^\varphi$*, *Opp$_\mathcal{C}^\varphi$* and *Ind$_\mathcal{C}^\varphi$* form a partition of $\mathcal{C}$, for any $\varphi \in \mathcal{L}^{\mathcal{A}}$. Note that these three sets do not depend on whether φ is strict or not. We may abbreviate *Supp$_\mathcal{C}^\varphi$* to *Supp$^\varphi$*, and similarly for *Opp$_\mathcal{C}^\varphi$* and *Ind$_\mathcal{C}^\varphi$*. *Supp$^\varphi$* are the evaluations that support φ; *Opp$^\varphi$* are the evaluations that oppose φ. *Ind$^\varphi$* are the other evaluations, that are indifferent regarding φ. For a model H to satisfy φ it is necessary that no evaluation that opposes φ appears before all evaluations that support φ. More precisely, we have the following:

Lemma 5. *Let H be an element of $\mathcal{C}(1)$, i.e., a sequence of different elements of $\mathcal{C}$. For strict φ, $H \models \varphi$ if and only if an element of Supp$_\mathcal{C}^\varphi$ appears in H which appears before any (if there are any) element in Opp$_\mathcal{C}^\varphi$ that appears. For non-strict φ, $H \models \varphi$ if and only if an element of Supp$_\mathcal{C}^\varphi$ appears in H before any element in Opp$_\mathcal{C}^\varphi$ appears, or no element of Opp$_\mathcal{C}^\varphi$ appears in H (i.e., $\sigma(H) \cap Opp_\mathcal{C}^\varphi = \emptyset$).*

Proof: Let $H = (c_1, \ldots c_k)$ be a $\mathcal{C}(1)$-model. Suppose that φ is a strict statement. Then $H \models \varphi$, i.e., $\alpha_\varphi \prec_H \beta_\varphi$, if and only if there exists some $i \in \{1, \ldots, k\}$ such that $\{c_1, \ldots c_{i-1}\} \subseteq Ind^\varphi$ and $c_i \in Supp_\mathcal{C}^\varphi$, which is if and only if an element of *Supp$_\mathcal{C}^\varphi$* appears in H before any element in *Opp$_\mathcal{C}^\varphi$* appears.

Now suppose that φ is a non-strict statement. Then $H \models \varphi$, i.e., $\alpha_\varphi \preccurlyeq_H \beta_\varphi$, if and only if either (i) for all $i = 1, \ldots, k$, $\alpha \equiv_{c_i} \beta$; or (ii) there exists some $i \in \{1, \ldots, k\}$ such that $\alpha \prec_{c_i} \beta$ and for all j such that $1 \leq j < i$, $\alpha \equiv_{c_j} \beta$. (i) holds if and only if $\sigma(H) \subseteq Ind^\varphi$, i.e., no element of *Supp$_\mathcal{C}^\varphi$* or *Opp$_\mathcal{C}^\varphi$* appears in H. (ii) holds if and only if an element of *Supp$_\mathcal{C}^\varphi$* appears in H before any element in *Opp$_\mathcal{C}^\varphi$* appears, and some element of *Supp$_\mathcal{C}^\varphi$* appears in H. Thus, $H \models \varphi$ holds if and only if either no element in *Opp$_\mathcal{C}^\varphi$* appears in H or some element of *Supp$_\mathcal{C}^\varphi$* appears in H and the first such element appears before any element in *Opp$_\mathcal{C}^\varphi$* appears. □

The following defines *inconsistency bases*, which are concerned with evaluations that cannot appear in any model satisfying the set of preference statements Γ (see Proposition 3 below). They are a valuable tool in understanding the structure of the set of satisfying models (see e.g., Proposition 8 below).

Definition 1. *Let $\Gamma \subseteq \mathcal{L}^{\mathcal{A}}$, and let $\mathcal{C}$ be a set of $\mathcal{A}$-evaluations. We say that $(\Gamma', \mathcal{C}')$ is an inconsistency base for $(\Gamma, \mathcal{C})$ if $\Gamma' \subseteq \Gamma$, and $\mathcal{C}' \subseteq \mathcal{C}$, and*

(i) for all $\varphi \in \Gamma'$, $Supp_\mathcal{C}^\varphi \cup Opp_\mathcal{C}^\varphi \subseteq C'$ (and thus $\mathcal{C} - C' \subseteq Ind_\mathcal{C}^\varphi$); and

(ii) for all $c \in C'$, there exists $\varphi \in \Gamma'$ such that $Opp_\mathcal{C}^\varphi \ni c$.

Thus, for all $\varphi \in \Gamma'$, the set C' contains all evaluations that are not indifferent regarding φ, and for all $c \in C'$ there is some element of Γ' that is opposed by c.

Example 5. *Consider evaluations $\mathcal{C} = \{e, f, g, h\}$ with values for alternatives α, β, γ and δ as in the following table.*

	α	β	γ	δ
e	2	2	3	3
f	0	3	1	1
g	0	2	2	0
h	1	1	3	2

Consider the strict preference statement $\varphi_1 : \alpha < \beta$, and the non-strict preference statements $\varphi_2 : \beta \leq \gamma$, $\varphi_3 : \gamma \leq \delta$. Let $\Gamma = \{\varphi_1, \varphi_2, \varphi_3\}$. Then, $Opp_\mathcal{C}^{\varphi_1} = \emptyset$, $Supp_\mathcal{C}^{\varphi_1} = \{f, g\}$ and $Ind_\mathcal{C}^{\varphi_1} = \{e, h\}$. Similarly, $Opp_\mathcal{C}^{\varphi_2} = \{f\}$, $Supp_\mathcal{C}^{\varphi_2} = \{e, h\}$ and $Ind_\mathcal{C}^{\varphi_2} = \{g\}$. For φ_3, $Opp_\mathcal{C}^{\varphi_3} = \{g, h\}$, $Supp_\mathcal{C}^{\varphi_3} = \emptyset$ and $Ind_\mathcal{C}^{\varphi_3} = \{e, f\}$.

The HCLP model (e, f) satisfies Γ. As stated in Lemma 5, the evaluation $e \in Supp_\mathcal{C}^{\varphi_2}$ precedes the only element f in $Opp_\mathcal{C}^{\varphi_2}$. The tuple $(\Gamma', C') = (\{\varphi_3\}, \{g, h\})$ is an inconsistency base of $(\Gamma, \mathcal{C})$. Condition (i) of Definition 1 is satisfied by $Supp_\mathcal{C}^{\varphi_3} \cup Opp_\mathcal{C}^{\varphi_3} = \{g, h\} \subseteq C'$. Since for $g, h \in C'$, $g \in Opp_\mathcal{C}^{\varphi_3}$ and $h \in Opp_\mathcal{C}^{\varphi_3}$, condition (ii) is satisfied as well.

The following result motivates the definition of inconsistency bases, showing that no model of Γ can involve any element of C', and that if Γ' contains a strict element then Γ is $\mathcal{C}(1)$-inconsistent.

Proposition 3. *Let (Γ', C') be an inconsistency base for $(\Gamma, \mathcal{C})$. Let H be an element of $\mathcal{C}(1)$. If $H \models \Gamma'$ then $C' \cap \sigma(H) = \emptyset$ and for any $\varphi \in \Gamma'$, $\alpha_\varphi \equiv_H \beta_\varphi$, so $H \not\models \alpha_\varphi < \beta_\varphi$. In particular, no $\mathcal{C}(1)$ model of Γ can involve any element of C'. Also, if Γ is $\mathcal{C}(1)$-consistent then Γ' contains no strict preference statements.*

Proof: Let (Γ', C') be an inconsistency base for $(\Gamma, \mathcal{C})$. Let $H = (c_1, \ldots c_k)$ be an element of $\mathcal{C}(1)$ with $H \models \Gamma'$. Suppose H contains some element in C' and let c_i be the element in $C' \cap \sigma(H)$ with the smallest index. By Definition 1(ii), there exists $\varphi \in \Gamma'$ such that $Opp_\mathcal{C}^\varphi \ni c_i$. Furthermore, since $c_j \notin C'$ for all $1 \leq j < i$, Definition 1(i) implies $c_j \in Ind_\mathcal{C}^\varphi$.

But then, an evaluation that opposes φ appears before all evaluations that support φ. By Lemma 5, this is a contradiction to $H \models \Gamma'$; hence we must have $C' \cap \sigma(H) = \emptyset$. Also, for all $\varphi \in \Gamma'$, $\sigma(H) \subseteq \mathcal{C} - C' \subseteq Ind_\mathcal{C}^\varphi$ by Definition 1(i). Therefore, for any $\varphi \in \Gamma'$, $\alpha_\varphi \equiv_H \beta_\varphi$, and thus $H \not\models \alpha_\varphi < \beta_\varphi$. Since $H \models \Gamma'$, this implies that Γ' contains no strict elements. The last parts follow from the fact that Γ' is a subset of Γ, so if $H \models \Gamma$ then $H \models \Gamma'$. □

We next give a small technical lemma that will be useful later. In particular, part (i) will be used in proving compactness of preference inference.

Lemma 6. *Assume that (Γ', C') is an inconsistency base for $(\Gamma, \mathcal{C})$. Then the following hold.*

(i) *There exists a finite set $\Gamma'' \subseteq \Gamma$ such that (Γ'', C') is an inconsistency base for $(\Gamma, \mathcal{C})$, and if Γ' contains a strict statement then Γ'' does also.*

(ii) *For any Δ such that $\Gamma' \subseteq \Delta \subseteq \Gamma$, (Γ', C') is an inconsistency base for $(\Delta, \mathcal{C})$.*

Proof: (i): By condition (ii) of the definition of an inconsistency base, for each $c \in C'$, there exists $\varphi_c \in \Gamma'$ such that $Opp_\mathcal{C}^{\varphi_c} \ni c$. If Γ' contains a strict statement ψ then let $\Gamma'' = \{\psi\} \cup \{\varphi_c : c \in C'\}$; else let $\Gamma'' = \{\varphi_c : c \in C'\}$. Because $\mathcal{C}$ is finite, Γ'' is finite. The definition implies that (Γ'', C') is an inconsistency base for $(\Gamma, \mathcal{C})$.

Part (ii) follows immediately from Definition 1, since conditions (i) and (ii) of the definition do not directly refer to Γ, but just to Γ', which is a subset of Γ. □

We will show there is, in a natural sense, a unique maximal inconsistency base for $(\Gamma, \mathcal{C})$.

For inconsistency bases (Γ_1, C_1) and (Γ_2, C_2) for $(\Gamma, \mathcal{C})$, define $(\Gamma_1, C_1) \cup (\Gamma_2, C_2)$ to be $(\Gamma_1 \cup \Gamma_2, C_1 \cup C_2)$. More generally, for inconsistency bases (Γ_i, C_i), $i \in I$, we define $\cup_{i \in I}(\Gamma_i, C_i)$ to be $(\cup_{i \in I}\Gamma_i, \cup_{i \in I}C_i)$, which can be easily shown to be an inconsistency base.

Lemma 7. *Suppose, for some (finite or infinite) non-empty index set I, and for all $i \in I$, that (Γ_i, C_i) is an inconsistency base. Then $\cup_{i \in I}(\Gamma_i, C_i)$ is an inconsistency base.*

Proof: For all $i \in I$, by Definition 1(i), for all $\varphi \in \Gamma_i$, $Supp_\mathcal{C}^\varphi \cup Opp_\mathcal{C}^\varphi \subseteq C_i$; thus, for all $\varphi \in \cup_{i \in I}\Gamma_i$, $Supp_\mathcal{C}^\varphi \cup Opp_\mathcal{C}^\varphi \subseteq \cup_{i \in I}C_i$. This proves condition (i). To prove condition (ii): for all $i \in I$, by Definition 1(ii), for all $c \in C_i$, there exists $\varphi \in \Gamma_i$ such that $Opp_\mathcal{C}^\varphi \ni c$. Thus, for all $c \in \cup_{i \in I}C_i$, there exists $\varphi \in \cup_{i \in I}\Gamma_i$ such that $Opp_\mathcal{C}^\varphi \ni c$. □

Define $MIB(\Gamma, \mathcal{C})$, the *maximal inconsistency base for* $(\Gamma, \mathcal{C})$, to be the union of all inconsistency bases for $(\Gamma, \mathcal{C})$, i.e., $\bigcup \{(\Gamma', C') \in \mathcal{I}\}$, where $\mathcal{I}$ is the set of inconsistency bases for $(\Gamma, \mathcal{C})$. This is well-defined, because $\mathcal{I}$ is non-empty, since it always contains the tuple $(\emptyset, \emptyset)$.

The next result states that $MIB(\Gamma, \mathcal{C})$ is an inconsistency base for $(\Gamma, \mathcal{C})$.

Proposition 4. *$MIB(\Gamma, \mathcal{C})$ is an inconsistency base for $(\Gamma, \mathcal{C})$, which is maximal in the following sense: if (Γ_1, C_1) is an inconsistency base for $(\Gamma, \mathcal{C})$ then $\Gamma_1 \subseteq \Gamma^\perp$ and $C_1 \subseteq C^\perp$, where $MIB(\Gamma, \mathcal{C}) = (\Gamma^\perp, C^\perp)$.*

Proof: By Lemma 7, the union of an arbitrary set of inconsistency bases is an inconsistency base. Consequently, $MIB(\Gamma, \mathcal{C})$ is an inconsistency base. Let $MIB(\Gamma, \mathcal{C}) = (\Gamma^\perp, C^\perp)$. The definition immediately implies that if (Γ_1, C_1) is an inconsistency base for $(\Gamma, \mathcal{C})$, then $\Gamma_1 \subseteq \Gamma^\perp$ and $C_1 \subseteq C^\perp$. □

By Proposition 3, if Γ is $\mathcal{C}(1)$-consistent then $\Gamma^\perp$ contains no strict elements, proving the next result. The converse also holds—see Corollary 1.

Proposition 5. *Suppose that Γ is $\mathcal{C}(1)$-consistent, i.e., there exists a $\mathcal{C}(1)$ model of Γ. Then for any inconsistency base (Γ', C') of $(\Gamma, \mathcal{C})$, $\Gamma' \cap \mathcal{L}_<^A = \emptyset$. In particular, if $MIB(\Gamma, \mathcal{C}) = (\Gamma^\perp, C^\perp)$ then $\Gamma^\perp \cap \mathcal{L}_<^A = \emptyset$.*

Example 6. *Consider the HCLP structure and preference statements as in Example 5. The only inconsistency bases of $(\Gamma, \mathcal{C})$ are $(\emptyset, \emptyset)$ and $(\{\varphi_3\}, \{g, h\})$. Thus, $(\{\varphi_3\}, \{g, h\})$ is the maximal inconsistency base $MIB(\Gamma, \mathcal{C})$ and does not contain any strict statements of Γ.*

In the following sections, it will be important to consider models extending other models.

Definition 2. *For $H, H' \in \mathcal{C}(1)$, write H as $(c_1, \ldots, c_k)$ and $H' = (c'_1, \ldots, c'_l)$. we say that H' extends H if $l > k$ and for all $j = 1, \ldots, k$, $c'_j = c_j$.*

Lemma 8. *Suppose that $H, H' \in \mathcal{C}(1)$ and that H' extends H. Then,*

(i) *If $H \models \alpha < \beta$ then $H' \models \alpha < \beta$.*

(ii) *If $H' \models \alpha \leq \beta$ then $H \models \alpha \leq \beta$.*

Proof: (i) Suppose that $H \models \alpha < \beta$, so that $\alpha \prec_H \beta$. Write H as $(c_1, \ldots, c_k)$. For some i, $c_i(\alpha) \neq c_i(\beta)$; and let i be minimal such that $c_i(\alpha) \neq c_i(\beta)$. Since $\alpha \prec_H \beta$, we have $c_i(\alpha) < c_i(\beta)$. Because, H' extends H, this implies that $\alpha \prec_{H'} \beta$, i.e., $H' \models \alpha < \beta$.

(ii) Suppose that $H' \models \alpha \leq \beta$. Then $H' \not\models \beta < \alpha$, by Lemma 1. Part (i) implies that $H \not\models \beta < \alpha$, and thus $H \models \alpha \leq \beta$, using Lemma 1 again. □

4.2 Towards a polynomial algorithm for consistency and deduction

Throughout this section we consider a set $\Gamma \subseteq \mathcal{L}^{\mathcal{A}}$ of input preference statements, and a set $\mathcal{C}$ of $\mathcal{A}$-evaluations.

Define $Opp_\Gamma(c)$ (usually abbreviated to $Opp(c)$) to be the set of elements opposed by c, i.e., $\varphi \in \Gamma$ such that $c(\alpha_\varphi) > c(\beta_\varphi)$, and define $Supp_\Gamma(c)$ (abbreviated to $Supp(c)$) to be the set of elements φ of Γ supported by c, (i.e., $c(\alpha_\varphi) < c(\beta_\varphi)$). For for $C' \subseteq \mathcal{C}$, we define $Supp_\Gamma(C')$ to be the elements of Γ that are supported by some element of C', i.e., $Supp(C') = \bigcup_{c \in C'} Supp(c)$. Also, for sequence of evaluations $(c_1, \ldots, c_k)$, we define $Supp(c_1, \ldots, c_k)$ to be $\bigcup_{i=1}^{k} Supp(c_i)$, which equals $Supp(\{c_1, \ldots, c_k\})$.

We thus have $\varphi \in Supp(c) \iff c(\alpha_\varphi) < c(\beta_\varphi) \iff c \in Supp^\varphi$; and $\varphi \in Opp(c) \iff c(\alpha_\varphi) > c(\beta_\varphi) \iff c \in Opp^\varphi$.

$\Gamma^{(\leq)}$, the non-strict version of Γ: It turns out to be helpful to consider a non-strict version of Γ; we define $\Gamma^{(\leq)}$ to be $\{\alpha_\varphi \leq \beta_\varphi : \varphi \in \Gamma\}$, i.e., Γ where the strict statements are replaced by corresponding non-strict statements. Clearly, if $H \models \Gamma$ then $H \models \Gamma^{(\leq)}$ (since $H \models \alpha < \beta$ implies $H \models \alpha \leq \beta$).

The next lemma follows immediately, since the definition of maximal inconsistency base does not depend on whether elements of Γ are strict or not.

Lemma 9. *For any Γ and $\mathcal{C}$, $MIB(\Gamma^{(\leq)}, \mathcal{C}) = MIB(\Gamma, \mathcal{C})$.*

In order to determine the consistency of set of preference statements Γ, we want a method for generating a model $H \in \mathcal{C}(1)$ satisfying Γ. (Determining (non-)inference can be similarly performed by generating a model satisfying $\Gamma \cup \{\neg\varphi\}$, using Proposition 1.) A necessary condition for $H \models \Gamma$ is $H \models \Gamma^{(\leq)}$. There is a simple necessary and sufficient condition for $H \models \Gamma^{(\leq)}$, where $H = (c_1, \ldots, c_k)$, which is that every $\varphi \in \Gamma$ that is opposed by c_j is supported by some earlier element in the sequence (see Proposition 6). This condition allows one to easily incrementally grow models of $\Gamma^{(\leq)}$, until one has a maximal model of $\Gamma^{(\leq)}$. We only need to consider maximal models because if a model H of $\Gamma^{(\leq)}$ satisfies Γ then any maximal model of $\Gamma^{(\leq)}$ extending H satisfies Γ (see Lemma 11). The results about maximal inconsistency bases allow us to show (Theorem 2) that if Γ is consistent then any maximal model of $\Gamma^{(\leq)}$ satisfies Γ, so to determine consistency of Γ we just need to generate any maximal model of $\Gamma^{(\leq)}$, which can be done in a straight-forward iterative way. This is the basis of the algorithm.

4.2.1 Γ-allowed sequences, i.e., models of $\Gamma^{(\leq)}$

We define the notion of Γ-allowed sequence, which turns out to be the same as a model of $\Gamma^{(\leq)}$ (see Proposition 6), and derive important properties (Proposition 7), which are useful for deriving the main results about maximal Γ-allowed sequences in Section 4.2.2.

Define $\textit{Next}_\Gamma(C')$ to be the set of all $c \in \mathcal{C} - C'$ such that $\textit{Opp}(c) \subseteq \textit{Supp}(C')$, i.e., the set of $c \in \mathcal{C} - C'$ that only oppose elements in Γ that are supported by elements of C'. The following result gives an equivalent condition for $c \in \textit{Next}_\Gamma(C')$.

Lemma 10. *Consider any $c \in \mathcal{C}$. Then, $c \in \textit{Next}_\Gamma(C')$, i.e., $\textit{Opp}(c) \subseteq \textit{Supp}(C')$, if and only if for all $\varphi \in \Gamma - \textit{Supp}(C')$, $c \in \textit{Supp}^\varphi \cup \textit{Ind}^\varphi$.*

Proof: Suppose first that $\textit{Opp}(c) \subseteq \textit{Supp}(C')$, and consider any $\varphi \in \Gamma - \textit{Supp}(C')$. Since $\varphi \notin \textit{Supp}(C')$, then $\varphi \notin \textit{Opp}(c)$, and thus, $c \notin \textit{Opp}^\varphi$. This implies that $c \in \textit{Supp}^\varphi \cup \textit{Ind}^\varphi$.

Conversely, suppose that for all $\varphi \in \Gamma - \textit{Supp}(C')$, $c \in \textit{Supp}^\varphi \cup \textit{Ind}^\varphi$. Consider any $\varphi \in \textit{Opp}(c)$. Then $c \in \textit{Opp}^\varphi$ and so $c \notin \textit{Supp}^\varphi \cup \textit{Ind}^\varphi$, and therefore, $\varphi \in \textit{Supp}(C')$. □

Consider an arbitrary sequence $H = (c_1, \ldots, c_k)$ of elements of $\mathcal{C}$. Let us say that H is a Γ-*allowed sequence (of $\mathcal{C}$)* if for all $j = 1, \ldots, k$, $c_j \in \textit{Next}(\{c_1, \ldots, c_{j-1}\})$, i.e., $\textit{Opp}(c_j) \subseteq \textit{Supp}(\{c_1, \ldots, c_{j-1}\})$. These turn out to be just models of $\Gamma^{(\leq)}$.

Example 7. *Consider the HCLP structure as in Example 5 and preference statements $\Gamma = \{\varphi_1, \varphi_2\}$ with $\varphi_1 : \alpha < \beta$ and $\varphi_2 : \beta \leq \gamma$. Then $H = (h, f, e)$ is a Γ-allowed sequence since:*

- $e \in \textit{Next}(\{h, f\})$, *i.e.,* $\textit{Opp}(e) = \emptyset \subseteq \textit{Supp}(\{h, f\}) = \{\varphi_1, \varphi_2\}$.
- $f \in \textit{Next}(\{h\})$, *i.e.,* $\textit{Opp}(f) = \{\varphi_2\} \subseteq \textit{Supp}(\{h\}) = \{\varphi_2\}$.
- $h \in \textit{Next}(\emptyset)$, *i.e.,* $\textit{Opp}(h) = \emptyset \subseteq \textit{Supp}(\emptyset) = \emptyset$.

H satisfies both preference statements in Γ.

Proposition 6. *Consider an arbitrary sequence $H = (c_1, \ldots, c_k)$ of elements of $\mathcal{C}$. Then, $H \models \Gamma^{(\leq)}$ if and only H is a Γ-allowed sequence.*

Proof: Suppose that $H \not\models \Gamma^{(\leq)}$, so there exists some $\varphi \in \Gamma$ such that $H \not\models \alpha_\varphi \leq \beta_\varphi$. If all elements c_j of H were indifferent to φ (i.e., $c_j(\alpha_\varphi) = c_j(\beta_\varphi)$) then we would have $H \models \alpha_\varphi \leq \beta_\varphi$. Thus, some element c_j in H is not indifferent to φ; let c_i be the first such element in H. If it were the case that $c_i(\alpha_\varphi) < c_i(\beta_\varphi)$ then we would have $H \models \alpha_\varphi \leq \beta_\varphi$, so we must have $c_i(\alpha_\varphi) > c_i(\beta_\varphi)$, and thus, $\varphi \in \textit{Opp}(c_i)$. Now, $\varphi \notin \textit{Supp}(\{c_1, \ldots, c_{i-1}\})$, since $c_j(\alpha_\varphi) = c_j(\beta_\varphi)$ for all $j < i$, and hence, $\textit{Opp}(c_i) \not\subseteq \textit{Supp}(\{c_1, \ldots, c_{i-1}\})$. This shows that $c_i \notin \textit{Next}(\{c_1, \ldots, c_{i-1}\})$, and so H is not a Γ-allowed sequence.

Conversely, suppose that for some $j \in \{1, \ldots, k\}$, $c_j \notin \textit{Next}(\{c_1, \ldots, c_{j-1}\})$, and let c_i be the first such c_j. Then for all $j < i$, $c_j \in \textit{Next}(\{c_1, \ldots, c_{j-1}\})$. Since $c_i \notin \textit{Next}(\{c_1, \ldots, c_{i-1}\})$, there exists some $\varphi \in \Gamma - \textit{Supp}(\{c_1, \ldots, c_{i-1}\})$ such that $\varphi \in$

$Opp(c_i)$. so that $c_i(\alpha_\varphi) > c_i(\beta_\varphi)$. Let j be minimal such that $c_j(\alpha_\varphi) \neq c_j(\beta_\varphi)$. Since $\varphi \notin Supp(\{c_1, \ldots, c_{i-1}\})$, we do not have $c_j(\alpha_\varphi) < c_j(\beta_\varphi)$, so we must have $c_j(\alpha_\varphi) > c_j(\beta_\varphi)$. This implies that $H \not\models \alpha_\varphi \leq \beta_\varphi$, where $\alpha_\varphi \leq \beta_\varphi$ is an element of $\Gamma^{(\leq)}$, and thus $H \not\models \Gamma^{(\leq)}$. □

We also have the following property of Γ-allowed sequences.

Proposition 7. *Suppose that H is a Γ-allowed sequence. Then, for all $\varphi \in Supp(H)$, $H \models \alpha_\varphi < \beta_\varphi$, and for all $\varphi \in \Gamma - Supp(H)$, $\alpha_\varphi \equiv_H \beta_\varphi$, so, in particular $H \not\models \alpha_\varphi < \beta_\varphi$. Thus, for $\varphi \in \Gamma$, we have $H \models \alpha_\varphi < \beta_\varphi$ if and only if $\varphi \in Supp(H)$. Also, $H \models \Gamma$ if and only if every strict element of Γ is in $Supp(H)$.*

Proof: First, consider any $\varphi \in Supp(H)$. Thus there exists $c_j \in \sigma(H)$ such that $c_j(\alpha_\varphi) < c_j(\beta_\varphi)$, so, in particular, $c_j(\alpha_\varphi) \neq c_j(\beta_\varphi)$. Let i be minimal such that $c_i(\alpha_\varphi) \neq c_i(\beta_\varphi)$. Proposition 6 implies that $H \models \alpha_\varphi \leq \beta_\varphi$, which implies that $c_i(\alpha_\varphi) \not> c_i(\beta_\varphi)$, and thus $c_i(\alpha_\varphi) < c_i(\beta_\varphi)$, proving that $H \models \alpha_\varphi < \beta_\varphi$.

Now, consider $\varphi \in \Gamma - Supp(H)$. If it were the case that there exists $c_j \in \sigma(H)$ such that $c_j(\alpha_\varphi) \neq c_j(\beta_\varphi)$, then the argument above implies that there exists i such that $c_i(\alpha_\varphi) < c_i(\beta_\varphi)$, and thus $\varphi \in Supp(H)$. Thus, for all $c_j \in \sigma(H)$, $c_j(\alpha_\varphi) = c_j(\beta_\varphi)$, and, hence, $\alpha_\varphi \equiv_H \beta_\varphi$.

For the last part, since, by Proposition 6, $H \models \Gamma^{(\leq)}$, we have: $H \models \Gamma$ if and only if for every strict element φ of Γ, $H \models \alpha_\varphi < \beta_\varphi$, i.e., $\varphi \in Supp(H)$. □

4.2.2 Maximal Γ-allowed sequences, i.e., maximal models of $\Gamma^{(\leq)}$

We say that H is a maximal Γ-allowed sequence of $\mathcal{C}$ if H is a Γ-allowed sequence of $\mathcal{C}$ and no extension of H is a Γ-allowed sequence of $\mathcal{C}$, i.e., $Next(\sigma(H)) = \emptyset$. More generally, when talking about maximal models, with respect to some set of models $\mathcal{D}$, we mean maximality with respect to the extension relation, so a model in $\mathcal{D}$ is ($\mathcal{D}$-)maximal if there is no element of $\mathcal{D}$ that extends it.

Lemma 11. *Suppose that $H, H' \in \mathcal{C}$ and $H, H' \models \Gamma^{(\leq)}$, and that H' extends H. Then for all $\varphi \in \Gamma$, if $H \models \varphi$ then $H' \models \varphi$. In particular, if $H \models \Gamma$ then $H' \models \Gamma$.*

Proof: Assume that $H, H' \models \Gamma^{(\leq)}$, and H' extends H. Consider any $\varphi \in \Gamma$, and suppose that $H \models \varphi$. If φ is non-strict then $\varphi \in \Gamma^{(\leq)}$ and so $H' \models \varphi$. If φ is strict, then Lemma 8(i) implies that $H' \models \varphi$. □

We use this in proving the next result, which shows that if we are interested in finding models of Γ it is sufficient to only consider maximal Γ-allowed sequences, i.e., maximal models of $\Gamma^{(\leq)}$.

Lemma 12. *If H is a Γ-allowed sequence, then either H is a maximal Γ-allowed sequence or there exists a maximal Γ-allowed sequence H' that extends H. Then, for all $\varphi \in \Gamma$, if $H \models \varphi$ then $H' \models \varphi$. In particular, if $H \models \Gamma$ then $H' \models \Gamma$.*

Proof: The *extends* relation on the finite set of Γ-allowed sequences is transitive and acyclic. It follows that for any Γ-allowed sequence H there exists a maximal Γ-allowed sequence extending H. The last part follows from previous result, Lemma 11 (using the equivalence stated by Proposition 6). □

The following key lemma shows the close relationship between maximal Γ-allowed sequences and the maximal inconsistency base.

Lemma 13. *Suppose that H is a maximal Γ-allowed sequence. Then $(\Gamma - Supp(H), \mathcal{C} - \sigma(H))$ equals $MIB(\Gamma, \mathcal{C})$.*

Proof: We first check the two conditions in the definition of an inconsistency base (see Definition 1). Consider any element φ of $\Gamma - Supp(H)$. Proposition 7 implies that $\alpha_\varphi \equiv_H \beta_\varphi$, so that for all $c \in \sigma(H)$, $c(\alpha_\varphi) = c(\beta_\varphi)$, and so $\sigma(H) \subseteq Ind^\varphi$, showing that Condition (i) holds. Now, consider any evaluation c in $\mathcal{C} - \sigma(H)$. By definition of a maximal Γ-allowed sequence, $Next(\sigma(H)) = \emptyset$, so $c \notin Next(\sigma(H))$. Therefore, by Lemma 10, there exists $\varphi \in \Gamma - Supp(H)$ such that $c \notin Supp^\varphi \cup Ind^\varphi$, so $c \in Opp^\varphi$, showing that Condition (ii) of an inconsistency base holds.

Write $MIB(\Gamma, \mathcal{C})$ as $(\Gamma^\perp, \mathcal{C}^\perp)$. Thus, by definition, $\Gamma - Supp(H) \subseteq \Gamma^\perp$ and $\mathcal{C} - \sigma(H) \subseteq \mathcal{C}^\perp$. Proposition 6 implies that $H \models \Gamma^{(\leq)}$. Lemma 9 implies that $MIB(\Gamma^{(\leq)}, \mathcal{C}) = (\Gamma^\perp, \mathcal{C}^\perp)$. Proposition 3 then implies that $\mathcal{C}^\perp \cap \sigma(H) = \emptyset$, and so, $\mathcal{C} - \sigma(H) \supseteq \mathcal{C}^\perp$. Thus, $\mathcal{C} - \sigma(H) = \mathcal{C}^\perp$.

Consider any $\varphi \in \Gamma^\perp$. By definition of an inconsistency base, $\mathcal{C} - \mathcal{C}^\perp \subseteq Ind^\varphi$, i.e., $\sigma(H) \subseteq Ind^\varphi$, which implies $\alpha_\varphi \equiv_H \beta_\varphi$, and so, by Proposition 7, $\varphi \in \Gamma - Supp(H)$. Thus, $\Gamma^\perp \subseteq \Gamma - Supp(H)$, and hence, $\Gamma^\perp = \Gamma - Supp(H)$, completing the proof that $(\Gamma - Supp(H), \mathcal{C} - \sigma(H))$ equals $(\Gamma^\perp, \mathcal{C}^\perp)$. □

Different maximal Γ-allowed sequences satisfy the same subset of Γ and involve the same subset of $\mathcal{C}$:

Proposition 8. *Suppose that H is a maximal Γ-allowed sequence. Write $MIB(\Gamma, \mathcal{C})$ as $(\Gamma^\perp, \mathcal{C}^\perp)$. Then $\Gamma^\perp = \Gamma - Supp(H)$ and $\mathcal{C}^\perp = \mathcal{C} - \sigma(H)$. Thus, if H' is another maximal Γ-allowed sequence, then $\sigma(H') = \sigma(H)$ and $Supp(H') = Supp(H)$. Also, for all $\varphi \in \Gamma$, $H \models \varphi \iff H' \models \varphi$, which is if and only if φ is not a strict element of $\Gamma^\perp$. Hence, every maximal Γ-allowed sequence satisfies the same elements of Γ.*

Proof: By Lemma 13, $\Gamma^\perp = \Gamma - Supp(H)$ and $C^\perp = C - \sigma(H)$. For any maximal Γ-allowed sequence H', $\sigma(H') = C - C^\perp = \sigma(H)$, and $Supp(H') = \Gamma - \Gamma^\perp = Supp(H)$.

To prove the last part, suppose that $\varphi \in \Gamma$ is such that $H \not\models \varphi$. Proposition 6 implies that φ is strict. Proposition 7 implies that $\varphi \notin Supp(H)$ and thus $\varphi \in \Gamma^\perp$. Conversely, if φ is strict and $\varphi \in \Gamma^\perp$ then $\varphi \notin Supp(H)$, so $H \not\models \varphi$ by Proposition 7. We have shown, for $\varphi \in \Gamma$, that $H \models \varphi$ if and only if φ is not a strict element of $\Gamma^\perp$; the same argument applies to H', so $H \models \varphi \iff H' \models \varphi$. □

No model of $\Gamma^{(\leq)}$ satisfies any element of Γ that is not satisfied by a maximal Γ-allowed sequence H.

Proposition 9. *Consider any maximal Γ-allowed sequence H, and any $H' \in \mathcal{C}(1)$ such that $H' \models \Gamma^{(\leq)}$. For any $\varphi \in \Gamma$, if $H' \models \varphi$ then $H \models \varphi$.*

Proof: Suppose that $\varphi \in \Gamma$ and $H \not\models \varphi$, and so, by Proposition 7, φ is strict and $\varphi \in \Gamma - Supp(H)$. Consider any model $H' \models \Gamma^{(\leq)}$. By Proposition 6, H' is a Γ-allowed sequence. By Lemma 12, there exists some maximal Γ-allowed sequence H'' that extends or equals H'. We have $Supp(H') \subseteq Supp(H'')$. Proposition 8 implies that $Supp(H) = Supp(H'')$, so $\varphi \notin Supp(H')$. Since φ is strict, $H' \not\models \varphi$, again using Proposition 7. □

The theorem below shows that to test consistency, one just needs to generate a single maximal Γ-allowed sequence (i.e., maximal model of $\Gamma^{(\leq)}$), which can be easily done using an iterative algorithm.

Theorem 2. *Γ is $\mathcal{C}(1)$-consistent if and only if some maximal Γ-allowed sequence satisfies Γ, which is if and only if every maximal Γ-allowed sequence satisfies Γ.*

Proof: First assume that Γ is $\mathcal{C}(1)$-consistent, so there exists some HCLP model $H \in \mathcal{C}(1)$ such that $H \models \Gamma$. This trivially implies that $H \models \Gamma^{(\leq)}$ (since $H \models \alpha < \beta \Rightarrow H \models \alpha \leq \beta$), so by Proposition 6, H is a Γ-allowed sequence. By Lemma 12, there exists a maximal Γ-allowed sequence H' that extends or equals H, and $H' \models \Gamma$. We have proved that some maximal Γ-allowed sequence satisfies Γ. The converse is obvious: if some maximal Γ-allowed sequence satisfies Γ then Γ is $\mathcal{C}(1)$-consistent. The last part of Proposition 8 implies that some maximal Γ-allowed sequence satisfies Γ, if and only if every maximal Γ-allowed sequence satisfies Γ. □

This leads to a simple characterisation of $\mathcal{C}(1)$-consistency using the maximal inconsistency base: Γ is $\mathcal{C}(1)$-consistent if and only if no inconsistency base involves any strict element of Γ.

Corollary 1. *Write $MIB(\Gamma, \mathcal{C})$ as $(\Gamma^\perp, C^\perp)$. Γ is $\mathcal{C}(1)$-consistent if and only if $\Gamma^\perp \cap \mathcal{L}_<^A = \emptyset$, which is if and only if $\Gamma^\perp$ is $\mathcal{C}(1)$-consistent. If Γ is $\mathcal{C}(1)$-inconsistent then there exists a* finite set $\Gamma' \subseteq \Gamma^\perp$ such that Γ' is $\mathcal{C}(1)$-inconsistent, and $(\Gamma', C^\perp)$ is an inconsistency base for $(\Gamma, \mathcal{C})$.

Proof: Let $\Gamma_< = \Gamma \cap \mathcal{L}_<^\mathcal{A}$. First, suppose that Γ is $\mathcal{C}(1)$-consistent. Then, by Theorem 2, any maximal Γ-allowed sequence H satisfies Γ. By Proposition 7, $\Gamma_< \subseteq Supp(H)$, and thus, $\Gamma_< \subseteq \Gamma - \Gamma^\perp$, by Proposition 8. Hence, $\Gamma_< \cap \Gamma^\perp = \emptyset$, and so $\Gamma^\perp \cap \mathcal{L}_<^\mathcal{A} = \emptyset$.

Conversely, suppose that $\Gamma^\perp \cap \mathcal{L}_<^\mathcal{A} = \emptyset$. Proposition 8 implies that for any maximal Γ-allowed sequence H, $\Gamma - \Gamma^\perp = Supp(H)$ and thus, $\Gamma_< \subseteq Supp(H)$. Proposition 7 then implies that $H \models \Gamma$, and so Γ is $\mathcal{C}(1)$-consistent.

If Γ is $\mathcal{C}(1)$-consistent then $\Gamma^\perp$ is $\mathcal{C}(1)$-consistent, since $\Gamma^\perp \subseteq \Gamma$. Conversely, suppose that $\Gamma^\perp$ is $\mathcal{C}(1)$-consistent. Lemma 6 implies that $(\Gamma^\perp, C^\perp)$ is an inconsistency base for $(\Gamma^\perp, \mathcal{C})$. Proposition 5 implies that $\Gamma^\perp \cap \mathcal{L}_<^\mathcal{A} = \emptyset$, which by the first part, implies that Γ is $\mathcal{C}(1)$-consistent.

Now suppose that Γ is $\mathcal{C}(1)$-inconsistent. The first part implies that $\Gamma^\perp$ contains a strict statement. By Lemma 6(i), there exists finite $\Gamma' \subseteq \Gamma^\perp$ such that (Γ', C') is an inconsistency base for $(\Gamma, \mathcal{C})$, and Γ' contains a strict statement. By Lemma 6(ii), (Γ', C') is an inconsistency base for $(\Gamma', \mathcal{C})$, and thus, by Proposition 5, Γ' is $\mathcal{C}(1)$-inconsistent, since it contains a strict statement. □

The following result shows that this kind of preference inference is compact.

Corollary 2. *Consider any $\Gamma \subseteq \mathcal{L}^\mathcal{A}$ and $\varphi \in \mathcal{L}^\mathcal{A}$.*

(i) *If Γ is $\mathcal{C}(1)$-inconsistent then there exists finite $\Gamma' \subseteq \Gamma$ which is $\mathcal{C}(1)$-inconsistent.*

(ii) *If $\Gamma \models_{\mathcal{C}(1)} \varphi$ then there exists finite $\Gamma' \subseteq \Gamma$ such that $\Gamma' \models_{\mathcal{C}(1)} \varphi$.*

Proof: (i) Suppose that Γ is $\mathcal{C}(1)$-inconsistent. The last part of Corollary 1 implies that then there exists finite $\Gamma' \subseteq \Gamma$ which is $\mathcal{C}(1)$-inconsistent.

(ii) Suppose that $\Gamma \models_{\mathcal{C}(1)} \varphi$. Then $\Gamma \cup \{\neg\varphi\}$ is $\mathcal{C}(1)$-inconsistent, by Proposition 1. Part (i) implies that there exists finite $\mathcal{C}(1)$-inconsistent $\Delta \subseteq \Gamma \cup \{\neg\varphi\}$. If $\Delta \subseteq \Gamma$ then we can let $\Gamma' = \Delta$, since trivially $\Delta \models_{\mathcal{C}(1)} \varphi$. Otherwise, $\Delta \ni \varphi$, and we let $\Gamma' = \Delta - \{\varphi\}$. We have $\Gamma' \subseteq \Gamma$, and $\Gamma' \models_{\mathcal{C}(1)} \varphi$, again by Proposition 1. □

4.2.3 The algorithm

The idea behind the algorithm is to build up a maximal $\Gamma^{(\leq)}$-satisfying sequence by repeatedly adding evaluations to the end; suppose that we have picked a sequence of elements of $\mathcal{C}$, C' being the set picked so far. We need to choose next an evaluation c such that, if c

opposes some φ in Γ, then φ is supported by some evaluation in C' (or else the generated sequence will not satisfy φ).

H is initialised as the empty sequence () of evaluations. $H \leftarrow H + c$ means that evaluation c is added to the end of H.

Function *Cons-check*$(\Gamma, \mathcal{C})$

$H \leftarrow ()$
for $k = 1, \ldots, |\mathcal{C}|$ **do**
   if $\exists\, c \in \mathcal{C} - \sigma(H)$ such that $Opp(c) \subseteq Supp(H)$
     then choose some such c; $H \leftarrow H + c$
     else stop
   end for
return H

Note that at each stage, an element of $Next_\Gamma(\sigma(H))$ is chosen, so at each stage H is a Γ-allowed sequence. Also, the termination condition is equivalent to $Next_\Gamma(\sigma(H)) = \emptyset$, which implies that the returned H is a maximal Γ-allowed sequence.

The algorithm involves often non-unique choices. However, if we wish, the choosing of c can be done based on an ordering $c_1, \ldots, c_m$ of $\mathcal{C}$, where, if there exists more than one $c \in \mathcal{C} - \sigma(H)$ such that $Opp(c) \subseteq Supp(H)$, we choose the element c_i fulfilling this condition that has smallest index i. The algorithm then becomes deterministic, with a unique result following from the given inputs.

A straight-forward implementation runs in $O(|\Gamma||\mathcal{C}|^2)$ time; however, a more careful implementation runs in $O(|\Gamma||\mathcal{C}|)$ time, which we now describe. Let H_k be the HCLP model after the k-th iteration of the for-loop. In every iteration of the for-loop, we update sets $Opp_k^\Delta(c) = Opp(c) - Supp(H_k)$ and $Supp_k^\Delta(c) = Supp(c) - Supp(H_k)$ for all $c \in \mathcal{C} - \sigma(H_k)$. This costs us $O(|\mathcal{C} - \sigma(H_k)| \times |Supp(H_k) \setminus Supp(H_{k-1})|) = O(|\mathcal{C} - \sigma(H_k)| \times |Supp_{k-1}^\Delta(c_k)|)$ more time for every iteration k in which we add evaluation c_k to H_{k-1}. However, the choice of the next evaluation c_k can be performed in constant time by marking evaluations c with $Opp_{k-1}^\Delta(c) = \emptyset$. Suppose the algorithm stops after $1 \leq l \leq |\mathcal{C}|$ iterations. Since all $Supp_{k-1}^\Delta(c_k)$ are disjoint, $\sum_{k=1}^{l} |Supp_{k-1}^\Delta(c_k)| = |Supp(H_l)| \leq |\Gamma|$. Altogether, the running time is $O(\sum_{k=1}^{l} |\mathcal{C} - \sigma(H_k)| \times |Supp_{k-1}^\Delta(c_k)|) \leq O(|\mathcal{C}| \times \sum_{k=1}^{l} |Supp_{k-1}^\Delta(c_k)|)$, and thus the running time is $O(|\mathcal{C}| \times |\Gamma|)$.

Properties of the Algorithm

The algorithm will always generate an HCLP model satisfying Γ if Γ is $\mathcal{C}(1)$-consistent. It can also be used for computing the maximal inconsistency base. The following result sums up some properties related to the algorithm.

Theorem 3. *Let H be a sequence returned by the algorithm with inputs Γ and $\mathcal{C}$, and write $MIB(\Gamma, \mathcal{C})$ as $(\Gamma^\perp, C^\perp)$. Then $C^\perp = \mathcal{C} - \sigma(H)$ (i.e., the evaluations that don't appear in H), and $\Gamma^\perp = \Gamma - Supp(H)$. We have that $H \models \Gamma^{(\leq)}$. Also, Γ is $\mathcal{C}(1)$-consistent if and only if $Supp(H)$ contains all the strict elements of Γ, which is if and only if $\Gamma^\perp \cap \mathcal{L}_<^A = \emptyset$. If Γ is $\mathcal{C}(1)$-consistent then $H \models \Gamma$.*

Proof: By the construction of the algorithm, H is a maximal Γ-allowed sequence, as observed earlier. Proposition 8 implies that $C^\perp = \mathcal{C} - \sigma(H)$ and $\Gamma^\perp = \Gamma - Supp(H)$. By Proposition 6, we have $H \models \Gamma^{(\leq)}$. Corollary 1 implies that Γ is $\mathcal{C}(1)$-consistent if and only if $\Gamma^\perp \cap \mathcal{L}_<^A = \emptyset$. Theorem 2 implies that Γ is $\mathcal{C}(1)$-consistent if and only if $H \models \Gamma$. Proposition 7 implies that $H \models \Gamma$ if and only if $Supp(H)$ contains all the strict elements of Γ. □

The algorithm therefore determines $\mathcal{C}(1)$-consistency, and hence $\mathcal{C}(1)$-deduction (because of Proposition 1), in polynomial time, and also generates the maximal inconsistency base.

4.2.4 The case of inconsistent Γ

For the case when Γ is not $\mathcal{C}(1)$-consistent, the output H of the algorithm is a model which, in a sense, comes closest to satisfying Γ: it always satisfies $\Gamma^{(\leq)}$, the non-strict version of Γ, and if any model $H' \in \mathcal{C}(1)$ satisfies $\Gamma^{(\leq)}$ and any element φ of Γ, then H also satisfies φ.

Proposition 10. *Let H be a sequence returned by the algorithm with inputs Γ and $\mathcal{C}$, and suppose that $H' \in \mathcal{C}(1)$ is such that $H' \models \Gamma^{(\leq)}$. Then, for all $\varphi \in \Gamma$, if $H' \models \varphi$ then $H \models \varphi$.*

Proof: Since H is a maximal Γ-allowed sequence, we have (by Proposition 6) that $H \models \Gamma^{(\leq)}$. Suppose that $H' \in \mathcal{C}(1)$ is such that $H' \models \Gamma^{(\leq)}$. Proposition 9 implies that if $H' \models \varphi$ then $H \models \varphi$. □

These properties suggest the following way of reasoning with $\mathcal{C}(1)$-inconsistent Γ. Let us define Γ' to be equal to $(\Gamma - \Gamma^\perp) \cup \Gamma^{(\leq)}$. By Theorem 3, this is equal to $Supp(H) \cup \Gamma^{(\leq)}$, where H is a model generated by the algorithm, enabling easy computation of Γ'. Γ' is $\mathcal{C}(1)$-consistent, since it is satisfied by H. We might then (re-)define the (non-monotonic) deductions from $\mathcal{C}(1)$-inconsistent Γ to be the deductions from Γ'.

4.3 Strong consistency and max-model inference

In the set of models $\mathcal{C}(1)$, we allow models involving any subset of $\mathcal{C}$, the set of evaluations. We could alternatively consider a semantics where we only allow models H that involve all elements of $\mathcal{C}$, i.e., with $\sigma(H) = \mathcal{C}$.

Let $\mathcal{C}(1^*)$ be the set of elements H of $\mathcal{C}(1)$ with $\sigma(H) = \mathcal{C}$. Γ is defined to be *strongly $\mathcal{C}(1)$-consistent* if and only if there exists a model $H \in \mathcal{C}(1^*)$ such that $H \models \Gamma$. Let $MIB(\Gamma, \mathcal{C}) = (\Gamma^\perp, C^\perp)$. Proposition 3 implies that, if Γ is strongly $\mathcal{C}(1)$-consistent then $C^\perp$ is empty, and $\Gamma^\perp$ consists of all the elements of Γ that are indifferent to all of $\mathcal{C}$, i.e., the set of $\varphi \in \Gamma$ such that $c(\alpha_\varphi) = c(\beta_\varphi)$ for all $c \in \mathcal{C}$.

There is an associated preference inference based on this restricted set of models. We write $\Gamma \models_{\mathcal{C}(1^*)} \varphi$ if $H \models \varphi$ holds for every $H \in \mathcal{C}(1^*)$ such that $H \models \Gamma$.

This form of deduction can be expressed in terms of strong consistency, as the following result shows.

Lemma 14. *If Γ is strongly $\mathcal{C}(1)$-consistent then $\Gamma \models_{\mathcal{C}(1^*)} \varphi$ holds if and only if $\Gamma \cup \{\neg\varphi\}$ is not strongly $\mathcal{C}(1)$-consistent.*

Proof: First suppose that $\Gamma \cup \{\neg\varphi\}$ is strongly $\mathcal{C}(1)$-consistent. Then there exists $H \in \mathcal{C}(1)$ such that $H \models \Gamma \cup \{\neg\varphi\}$ and $\sigma(H) = \mathcal{C}$. Thus $H \models \Gamma$ and $H \not\models \varphi$ (using Lemma 1), showing that $\Gamma \not\models_{\mathcal{C}(1^*)} \varphi$.

Now suppose that $\Gamma \not\models_{\mathcal{C}(1^*)} \varphi$. Then there exists $H \in \mathcal{C}(1)$ such that $H \models \Gamma$ and $\sigma(H) = \mathcal{C}$ and $H \not\models \varphi$. Then $H \models \Gamma \cup \{\neg\varphi\}$ (again using Lemma 1), so $\Gamma \cup \{\neg\varphi\}$ is strongly $\mathcal{C}(1)$-consistent. $\square$

In the next section we will consider a related (and, in a sense, more general) form of preference inference, where we only consider maximal models.

4.3.1 Max-model inference

For $\Gamma \subseteq \mathcal{L}^{\mathcal{A}}$, let $\mathcal{M}_{\mathcal{C}(1)}^{\max}(\Gamma)$ be the set of maximal models within $\mathcal{C}(1)$ of Γ, i.e., the set of $H \in \mathcal{C}(1)$ such that $H \models \Gamma$, and for all $H' \in \mathcal{C}(1)$ extending H, $H' \not\models \Gamma$. We define the max-model inference relation $\models_{\mathcal{C}(1)}^{\max}$ by:

$\Gamma \models_{\mathcal{C}(1)}^{\max} \varphi$ if and only if $H \models \varphi$ for all $H \in \mathcal{M}_{\mathcal{C}(1)}^{\max}(\Gamma)$.

The following result shows that maximal models of Γ involve the same set of evaluations. It also shows that, if Γ is $\mathcal{C}(1)$-consistent, the maximal models are the same as the maximal Γ-allowed sequences discussed earlier.

Proposition 11. *Suppose that Γ is $\mathcal{C}(1)$-consistent. Then, for $H \in \mathcal{C}(1)$, we have $H \in \mathcal{M}_{\mathcal{C}(1)}^{\max}(\Gamma)$ if and only if H is a maximal Γ-allowed sequence in $\mathcal{C}$. Thus, for all $H, H' \in \mathcal{M}_{\mathcal{C}(1)}^{\max}(\Gamma)$, we have $\sigma(H) = \sigma(H') = \mathcal{C} - C^\perp$, where $MIB(\Gamma, \mathcal{C}) = (\Gamma^\perp, C^\perp)$.*

Proof: Consider any $H \in \mathcal{M}^{\max}_{\mathcal{C}(1)}(\Gamma)$. Since $H \models \Gamma$ we have $H \models \Gamma^{(\leq)}$, and so Proposition 6 implies that H is a Γ-allowed sequence. Suppose that H is not a maximal Γ-allowed sequence. Then, by Lemma 12, there exists a maximal Γ-allowed sequence H' extending H, and $H \models \Gamma$. This contradicts $H \in \mathcal{M}^{\max}_{\mathcal{C}(1)}(\Gamma)$.

Conversely, suppose that H is a maximal Γ-allowed sequence in $\mathcal{C}$. Theorem 2 implies that $H \models \Gamma$. To prove a contradiction, suppose that $H \notin \mathcal{M}^{\max}_{\mathcal{C}(1)}(\Gamma)$, so that there exists $H' \in \mathcal{M}^{\max}_{\mathcal{C}(1)}(\Gamma)$ with H' extending H. The argument above implies that H' is a maximal Γ-allowed sequence, which contradicts H being a maximal Γ-allowed sequence.

The last part follows from Proposition 8. $\square$

The next result shows that the same non-strict preference statements are inferred for the max-model inference relation $\models^{\max}_{\mathcal{C}(1)}$ as for the inference relation $\models_{\mathcal{C}(1)}$.

Proposition 12. *Consider any $\Gamma \subseteq \mathcal{L}^{\mathcal{A}}$, and any preference statement $\alpha \leq \beta$ in $\mathcal{L}^{\mathcal{A}}$.*

(i) *Γ is $\mathcal{C}(1)$-consistent if and only if $\mathcal{M}^{\max}_{\mathcal{C}(1)}(\Gamma) \neq \emptyset$.*

(ii) *$\Gamma \models^{\max}_{\mathcal{C}(1)} \alpha \leq \beta \iff \Gamma \models_{\mathcal{C}(1)} \alpha \leq \beta$.*

Proof: (i) follows easily: if Γ is $\mathcal{C}(1)$-consistent, then there exists some $H \in \mathcal{C}(1)$ with $H \models \Gamma$, so there exists $H' \in \mathcal{M}^{\max}_{\mathcal{C}(1)}(\Gamma)$ extending or equalling H. The converse is immediate: if there exists $H \in \mathcal{M}^{\max}_{\mathcal{C}(1)}(\Gamma)$ then $H \in \mathcal{C}(1)$ and $H \models \Gamma$, so Γ is $\mathcal{C}(1)$-consistent.

(ii) If Γ is not $\mathcal{C}(1)$-consistent then by part (i), $\mathcal{M}^{\max}_{\mathcal{C}(1)}(\Gamma) = \emptyset$, so $\Gamma \models^{\max}_{\mathcal{C}(1)} \alpha \leq \beta$ and $\Gamma \models_{\mathcal{C}(1)} \alpha \leq \beta$ both hold vacuously. Let us thus now assume that Γ is $\mathcal{C}(1)$-consistent.

$\Rightarrow$: Assume $\Gamma \models^{\max}_{\mathcal{C}(1)} \alpha \leq \beta$, and consider any $H \in \mathcal{C}(1)$ such that $H \models \Gamma$. We need to show that $H \models \alpha \leq \beta$. Since $H \models \Gamma$, we have $H \models \Gamma^{(\leq)}$, and so H is a Γ-allowed $\mathcal{C}$-sequence, by Proposition 6. Choose, by Lemma 12, any maximal Γ-allowed sequence H' extending or equalling H, and we have $H' \models \Gamma$. By, Proposition 11, $H' \in \mathcal{M}^{\max}_{\mathcal{C}(1)}(\Gamma)$. Then, $\Gamma \models^{\max}_{\mathcal{C}(1)} \alpha \leq \beta$ implies that $H' \models \alpha \leq \beta$. Lemma 8(ii) then implies that $H \models \alpha \leq \beta$.

$\Leftarrow$: Assume $\Gamma \models_{\mathcal{C}(1)} \alpha \leq \beta$, and consider any $H \in \mathcal{M}^{\max}_{\mathcal{C}(1)}(\Gamma)$. This implies that $H \in \mathcal{C}(1)$ and $H \models \Gamma$, so $H \models \alpha \leq \beta$ showing that $\Gamma \models^{\max}_{\mathcal{C}(1)} \alpha \leq \beta$. $\square$

We write $\Gamma \models_{\mathcal{C}(1)} \alpha \equiv \beta$ as an abbreviation of the conjunction of $\Gamma \models_{\mathcal{C}(1)} \alpha \leq \beta$ and $\Gamma \models_{\mathcal{C}(1)} \beta \leq \alpha$; and similarly for other inference relations. The last result can be used to prove that inferred equivalences are the same for max-model inference, and have a simple form.

Proposition 13. *Consider any $\mathcal{C}(1)$-consistent $\Gamma \subseteq \mathcal{L}^{\mathcal{A}}$, and any $\mathcal{C}$. Let $MIB(\Gamma, \mathcal{C})$ equal $(\Gamma^{\perp}, \mathcal{C}^{\perp})$. Consider any $\alpha, \beta \in \mathcal{A}$. Then, $\Gamma \models_{\mathcal{C}(1)} \alpha \equiv \beta$ if and only if $\Gamma \models^{\max}_{\mathcal{C}(1)} \alpha \equiv \beta$ if and only if for all $c \in \mathcal{C} - \mathcal{C}^{\perp}$, $c(\alpha) = c(\beta)$.*

Proof: First assume that $\Gamma \models_{\mathcal{C}(1)} \alpha \equiv \beta$. This trivially implies that $\Gamma \models_{\mathcal{C}(1)}^{\max} \alpha \equiv \beta$, since $\models_{\mathcal{C}(1)}^{\max} \subseteq \models_{\mathcal{C}(1)}$.

Now assume that $\Gamma \models_{\mathcal{C}(1)}^{\max} \alpha \equiv \beta$. Γ is $\mathcal{C}(1)$-consistent so $\mathcal{M}_{\mathcal{C}(1)}^{\max}(\Gamma) \neq \emptyset$, by Proposition 12(i). Consider any $H \in \mathcal{M}_{\mathcal{C}(1)}^{\max}(\Gamma)$. Then $\alpha \equiv_H \beta$, which implies that for all $c \in \sigma(H)$, $c(\alpha) = c(\beta)$, and thus, by Proposition 11, for all $c \in \mathcal{C} - \mathcal{C}^\perp$, $c(\alpha) = c(\beta)$.

Finally, let us assume that for all $c \in \mathcal{C} - \mathcal{C}^\perp$, $c(\alpha) = c(\beta)$. Consider any $H \in \mathcal{C}(1)$ such that $H \models \Gamma$. Proposition 3 implies that $\sigma(H) \cap \mathcal{C}^\perp = \emptyset$, i.e., $\sigma(H) \subseteq \mathcal{C} - \mathcal{C}^\perp$. So, for all $c \in \sigma(H)$, $c(\alpha) = c(\beta)$, and thus $\alpha \equiv_H \beta$, and hence $\Gamma \models_{\mathcal{C}(1)} \alpha \equiv \beta$. This completes the proof that the three statements are equivalent. $\square$

The following result shows that the strict inferences with $\models_{\mathcal{C}(1)}^{\max}$ are closely tied with the non-strict inferences.

Proposition 14. $\Gamma \models_{\mathcal{C}(1)}^{\max} \alpha \leq \beta$ *if and only if either* $\Gamma \models_{\mathcal{C}(1)}^{\max} \alpha \equiv \beta$ *or* $\Gamma \models_{\mathcal{C}(1)}^{\max} \alpha < \beta$. *Also, if Γ is $\mathcal{C}(1)$-consistent then $\Gamma \models_{\mathcal{C}(1)}^{\max} \alpha < \beta$ holds if and only if $\Gamma \models_{\mathcal{C}(1)}^{\max} \alpha \leq \beta$ and $\Gamma \not\models_{\mathcal{C}(1)}^{\max} \alpha \equiv \beta$.*

Proof: If Γ is not $\mathcal{C}(1)$-consistent then, by Proposition 12(i), $\mathcal{M}_{\mathcal{C}(1)}^{\max}(\Gamma) = \emptyset$, so $\Gamma \models_{\mathcal{C}(1)}^{\max} \alpha \leq \beta$ and $\Gamma \models_{\mathcal{C}(1)}^{\max} \alpha \equiv \beta$ (and $\Gamma \models_{\mathcal{C}(1)}^{\max} \alpha < \beta$) hold vacuously, and therefore the equivalence holds. Let us thus now assume that Γ is $\mathcal{C}(1)$-consistent. One direction holds easily: suppose that $\Gamma \models_{\mathcal{C}(1)}^{\max} \alpha \equiv \beta$ or $\Gamma \models_{\mathcal{C}(1)}^{\max} \alpha < \beta$, and consider any $H \in \mathcal{M}_{\mathcal{C}(1)}^{\max}(\Gamma)$. We have either $\alpha \equiv_H \beta$ or $H \models \alpha < \beta$, so either $\alpha \equiv_H \beta$ or $\alpha \prec_H \beta$, and thus $\alpha \preccurlyeq_H \beta$, and $H \models \alpha \leq \beta$, showing that $\Gamma \models_{\mathcal{C}(1)}^{\max} \alpha \leq \beta$.

Now, let us assume that $\Gamma \models_{\mathcal{C}(1)}^{\max} \alpha \leq \beta$, and that it is not the case that $\Gamma \models_{\mathcal{C}(1)}^{\max} \alpha \equiv \beta$. It is sufficient to show that $\Gamma \models_{\mathcal{C}(1)}^{\max} \alpha < \beta$. Consider any $H \in \mathcal{M}_{\mathcal{C}(1)}^{\max}(\Gamma)$. Since, $\Gamma \models_{\mathcal{C}(1)}^{\max} \alpha \leq \beta$, we have $H \models \alpha \leq \beta$. Proposition 13 implies that there exists $c \in \mathcal{C} - \mathcal{C}^\perp$ such that $c(\alpha) \neq c(\beta)$, where $MIB(\Gamma, \mathcal{C}) = (\Gamma^\perp, \mathcal{C}^\perp)$. By, Proposition 11, $\sigma(H) = \mathcal{C} - \mathcal{C}^\perp$, so there exists some $c \in \sigma(H)$ such that $c(\alpha) \neq c(\beta)$; let c be earliest such element of $\sigma(H)$. Since $H \models \alpha \leq \beta$, we have $c(\alpha) < c(\beta)$, so $H \models \alpha < \beta$. This shows that $\Gamma \models_{\mathcal{C}(1)}^{\max} \alpha < \beta$, as required.

Assume that Γ is $\mathcal{C}(1)$-consistent. Suppose that $\Gamma \models_{\mathcal{C}(1)}^{\max} \alpha < \beta$ holds. Then clearly, $\Gamma \models_{\mathcal{C}(1)}^{\max} \alpha \leq \beta$. Consider any $H \models \Gamma$. Then we have $\alpha \prec_H \beta$, so we do not have $\alpha \equiv_H \beta$, which implies that $\Gamma \models_{\mathcal{C}(1)} \alpha \equiv \beta$ does not hold. Conversely, suppose that $\Gamma \models_{\mathcal{C}(1)}^{\max} \alpha \leq \beta$ and $\Gamma \not\models_{\mathcal{C}(1)}^{\max} \alpha \equiv \beta$. The first part then implies that $\Gamma \models_{\mathcal{C}(1)}^{\max} \alpha < \beta$. $\square$

4.3.2 Properties of strong consistency and of the associated inference

The following result shows that the consequences of Γ with respect to $\models_{\mathcal{C}(1^*)}$ are the same as those with respect to $\models_{\mathcal{C}(1)}^{\max}$, when Γ is strongly $\mathcal{C}(1)$-consistent. (Of course, if Γ is not strongly $\mathcal{C}(1)$-consistent then all φ in $\mathcal{L}^A$ are consequences of $\models_{\mathcal{C}(1^*)}$.)

Lemma 15. *If Γ is strongly $\mathcal{C}(1)$-consistent then, for any $\varphi \in \mathcal{L}^A$, $\Gamma \models_{\mathcal{C}(1^*)} \varphi \iff \Gamma \models_{\mathcal{C}(1)}^{\max} \varphi$.*

Proof: Assume that Γ is strongly $\mathcal{C}(1)$-consistent, so there exists a model H' with $\sigma(H') = \mathcal{C}$. By definition of $\models_{\mathcal{C}(1^*)}$ and $\models_{\mathcal{C}(1)}^{\max}$ it is sufficient to show that $\mathcal{M}_{\mathcal{C}(1)}^{\max}(\Gamma)$ is equal to the set $\mathcal{H}$ of all $H \in \mathcal{C}(1)$ such that $H \models \Gamma$ and $\sigma(H) = \mathcal{C}$. It immediately follows that $\mathcal{M}_{\mathcal{C}(1)}^{\max}(\Gamma) \supseteq \mathcal{H}$. Conversely, consider any $H \in \mathcal{M}_{\mathcal{C}(1)}^{\max}(\Gamma)$. Since $H' \in \mathcal{H}$, we have $H' \in \mathcal{M}_{\mathcal{C}(1)}^{\max}(\Gamma)$. Proposition 11 implies that $\sigma(H) = \sigma(H') = \mathcal{C}$, proving that $H \in \mathcal{H}$. □

The next result shows that the non-strict $\models_{\mathcal{C}(1^*)}$ inferences are the same as the non-strict $\models_{\mathcal{C}(1)}$ inferences, and that (in contrast to the case of $\models_{\mathcal{C}(1)}$) the strict $\models_{\mathcal{C}(1^*)}$ inferences almost correspond with the non-strict ones. The result also implies that the algorithm in Section 4.2 can be used to efficiently determine the $\models_{\mathcal{C}(1^*)}$ inferences.

To illustrate the difference between the $\models_{\mathcal{C}(1)}$ inferences and the $\models_{\mathcal{C}(1^*)}$ inferences for the case of strict statements, consider some strongly $\mathcal{C}(1)$-consistent Γ which only includes non-strict statements. Then, for every strict preference statement $\alpha < \beta$, we will have $\Gamma \not\models_{\mathcal{C}(1)} \alpha < \beta$ since the empty sequence satisfies Γ but not $\alpha < \beta$. However, we will have $\Gamma \models_{\mathcal{C}(1^*)} \alpha < \beta$ if $\Gamma \models_{\mathcal{C}(1)} \alpha \leq \beta$ and $\Gamma \not\models_{\mathcal{C}(1)} \beta \leq \alpha$. For example, if Γ is just $\{\alpha \leq \beta\}$, where for some $c \in \mathcal{C}$, $c(\alpha) < c(\beta)$, then we will have $\Gamma \models_{\mathcal{C}(1^*)} \alpha < \beta$ but not $\Gamma \models_{\mathcal{C}(1)} \alpha < \beta$.

Proposition 15. *Let $MIB(\Gamma, \mathcal{C}) = (\Gamma^\perp, C^\perp)$. Γ is strongly $\mathcal{C}(1)$-consistent if and only if $C^\perp = \emptyset$ and $\Gamma \cap \mathcal{L}_<^A \subseteq \text{Supp}(\mathcal{C})$.*

Suppose that Γ is strongly $\mathcal{C}(1)$-consistent. Then,

(i) $\Gamma \models_{\mathcal{C}(1)} \alpha \leq \beta \iff \Gamma \models_{\mathcal{C}(1^*)} \alpha \leq \beta$;

(ii) $\Gamma \models_{\mathcal{C}(1^*)} \alpha \equiv \beta$ *if and only if α and β agree on all of $\mathcal{C}$, i.e., for all $c \in \mathcal{C}$, $c(\alpha) = c(\beta)$;*

(iii) $\Gamma \models_{\mathcal{C}(1^*)} \alpha < \beta$ *if and only if $\Gamma \models_{\mathcal{C}(1)} \alpha \leq \beta$ and α and β differ on some element of $\mathcal{C}$, i.e., there exists $c \in \mathcal{C}$ such that $c(\alpha) \neq c(\beta)$.*

Proof: First, suppose that Γ is strongly $\mathcal{C}(1)$-consistent. Then there exists $H' \in \mathcal{C}(1)$ such that $H' \models \Gamma$ and $\sigma(H') = \mathcal{C}$. Since $H' \models \Gamma^{(\leq)}$, by Proposition 6, H' is a Γ-allowed sequence. By Lemma 12, there exists a maximal Γ-allowed sequence H extending or equalling H', so, since $\sigma(H') = \mathcal{C}$, we must have $H = H'$. Proposition 8 implies that $C^\perp = \emptyset$ and $\Gamma^\perp = \Gamma - \text{Supp}(H) = \Gamma - \text{Supp}(\mathcal{C})$, and Corollary 1 shows then that $(\Gamma - \text{Supp}(\mathcal{C})) \cap \mathcal{L}_<^A = \emptyset$, which implies that $\Gamma \cap \mathcal{L}_<^A \subseteq \text{Supp}(\mathcal{C})$.

Conversely, suppose that $C^\perp = \emptyset$ and $\Gamma \cap \mathcal{L}^\mathcal{A}_< \subseteq Supp(\mathcal{C})$. Let H be a maximal Γ-allowed sequence. Proposition 8 implies that $\sigma(H) = \mathcal{C}$. Then $Supp(H) = Supp(\mathcal{C})$, and Proposition 7 implies that $H \models \Gamma$, showing that Γ is strongly $\mathcal{C}(1)$-consistent.

Now suppose that Γ is strongly $\mathcal{C}(1)$-consistent. Lemma 15 implies that for any $\varphi \in \mathcal{L}^\mathcal{A}$, $\Gamma \models_{\mathcal{C}(1^*)} \varphi \iff \Gamma \models^{\max}_{\mathcal{C}(1)} \varphi$. Part (i) then follows by Proposition 12(ii). Part (ii) follows from Proposition 13, using the fact that $C^\perp$ is empty. Part (iii) follows from part (ii) and Proposition 14. □

The next result shows that $\models_{\mathcal{C}(1)}$ inference is not affected if one removes the evaluations in the MIB.

Proposition 16. *Suppose that Γ is $\mathcal{C}(1)$-consistent, let $MIB(\Gamma, \mathcal{C}) = (\Gamma^\perp, C^\perp)$, and let $\mathcal{C}' = \mathcal{C} - C^\perp$. Then Γ is strongly $\mathcal{C}'(1)$-consistent, and $\Gamma \models_{\mathcal{C}(1)} \varphi$ if and only if $\Gamma \models_{\mathcal{C}'(1)} \varphi$.*

Proof: By Theorem 3, any output of the algorithm is in $\mathcal{C}'(1^*)$ and satisfies Γ. Thus Γ is strongly $\mathcal{C}'(1)$-consistent. Let $\mathcal{H}' = \{H \in \mathcal{C}'(1) : H \models \Gamma\}$ and $\mathcal{H} = \{H \in \mathcal{C}(1) : H \models \Gamma\}$. Then $\mathcal{H}' \subseteq \mathcal{H}$, because $\mathcal{C}'(1) \subseteq \mathcal{C}(1)$. By Proposition 3, for every $H \in \mathcal{H}$, we have $\sigma(H) \cap C^\perp = \emptyset$, and hence $H \in \mathcal{H}'$. Thus $\mathcal{H}' = \mathcal{H}$ and $\Gamma \models_{\mathcal{C}(1)} \varphi$ if and only if $\Gamma \models_{\mathcal{C}'(1)} \varphi$. □

4.4 Orderings on evaluations

The preference logic defined here is closely related to a logic based on disjunctive ordering statements. Given set of evaluations $\mathcal{C}$, we consider the set of statements $\mathcal{O}_\mathcal{C}$ of the form $C_1 < C_2$, and of $C_1 \leq C_2$, where C_1 and C_2 are disjoint subsets of $\mathcal{C}$.

We say that $H \models C_1 < C_2$ if some evaluation in C_1 appears in H before every element of C_2, that is, there exists some element of C_1 in H (i.e., $C_1 \cap \sigma(H) \neq \emptyset$) and the earliest element of $C_1 \cup C_2$ to appear in H is in C_1.

We say that $H \models C_1 \leq C_2$ if either $H \models C_1 < C_2$ or no element of C_1 or C_2 appears in H: $(C_1 \cup C_2) \cap \sigma(H) = \emptyset$. By Lemma 5 we have that
$$H \models \alpha_\varphi < \beta_\varphi \iff H \models Supp^\varphi_\mathcal{C} < Opp^\varphi_\mathcal{C},$$
and $H \models \alpha_\varphi \leq \beta_\varphi \iff H \models Supp^\varphi_\mathcal{C} \leq Opp^\varphi_\mathcal{C}$.

This shows that the language $\mathcal{O}_\mathcal{C}$ can express anything that can be expressed in $\mathcal{L}^\mathcal{A}$. It can be shown, conversely, that for any statement τ in $\mathcal{O}_\mathcal{C}$, one can define α_φ and β_φ, and the values of elements of $\mathcal{C}$ on these, such that for all $H \in \mathcal{C}(1)$, $H \models \tau$ if and only if $H \models \varphi$ (where φ is strict if and only if τ is strict). For instance, if τ is the statement $C_1 < C_2$, we can define $c(\alpha_\varphi) = 1$ for all $c \in C_2$, and $c(\alpha_\varphi) = 0$ for $c \in \mathcal{C} - C_2$; and define $c(\beta_\varphi) = 1$ for all $c \in C_1$, and $c(\beta_\varphi) = 0$ for $c \in \mathcal{C} - C_1$.

The algorithm adapts in the obvious way to the case where we have Γ consisting of (or including) elements in $\mathcal{O}_\mathcal{C}$. When viewed in this way, the algorithm can be seen as a simple extension of a topological sort algorithm; the standard case corresponds to when the ordering statements only involve singleton sets.

5 Proof theory for simple lexicographic inference

Preference inference has been defined semantically, and we have an efficient algorithm for the simple lexicographic case. From a logical perspective, it is natural to consider if we can construct an equivalent syntactical definition of inference via a proof theory; this can give another view of the assumptions being made by the logic. In this section we construct such a proof theory for preference inference based on simple lexicographic models, involving an axiom schema and a number of fairly simple inference rules. We consider a fixed set of evaluations $\mathcal{C}$ here, and we abbreviate $\models_{\mathcal{C}(1)}$ to just $\models$.

We make use of a form of Pareto (pointwise) ordering on alternatives, and we define a kind of addition and rescaling operation on alternatives and thus on preference statements.

We define the following pointwise (or weak Pareto) ordering on alternatives. For $\alpha, \beta \in \mathcal{A}$, $\alpha \preceq_{par} \beta \iff$ for all $c \in \mathcal{C}$, $c(\alpha) \leq c(\beta)$. We also define the Pareto Difference relation between elements of $\mathcal{L}^{\mathcal{A}}$. For $\psi, \theta \in \mathcal{L}^{\mathcal{A}}$, we say that $\psi \preceq_{parD} \theta$ holds if and only if (i) ψ and θ are either both strict or both non-strict; and (ii) for all $c \in \mathcal{C}$, $c(\beta_\psi) - c(\alpha_\psi) \leq c(\beta_\theta) - c(\alpha_\theta)$. Thus, if $\psi \preceq_{parD} \theta$ and $c(\alpha_\psi) \leq c(\beta_\psi)$ then $c(\alpha_\theta) \leq c(\beta_\theta)$. If $\psi \preceq_{parD} \theta$ and $H \models \psi$ then $H \models \theta$ (see Lemma 16(vi) below).

Pointwise multiplication of alternatives and preference statements: Let F be the set of functions from $\mathcal{C}$ to the strictly positive rational numbers. For $f \in F$, we define $\frac{1}{f} \in F$ in the obvious way, by, for $c \in \mathcal{C}$, $\frac{1}{f}(c) = \frac{1}{f(c)}$. Let f be an arbitrary element of F.

- For $\alpha, \gamma \in \mathcal{A}$, we say that $\alpha \doteq f\gamma$ if for all $c \in \mathcal{C}$, $c(\alpha) = f(c) \times c(\gamma)$ (where $\times$ is the standard multiplication).

- For $\varphi, \psi \in \mathcal{L}^{\mathcal{A}}$, we say that $\varphi \doteq f\psi$ if (i) $\alpha_\varphi \doteq f\alpha_\psi$ and $\beta_\varphi \doteq f\beta_\psi$, and (ii) φ is strict if and only if ψ is strict.

Note that if $\varphi \doteq f\psi$ then for all $c \in \mathcal{C}$, $c(\alpha_\varphi) \leq c(\beta_\varphi) \iff c(\alpha_\psi) \leq c(\beta_\psi)$. It is then easy to show that if $H \in \mathcal{C}(1)$ and $\varphi \doteq f\psi$ then $H \models \varphi$ if and only if $H \models \psi$: see Lemma 16(iv).

Addition of alternatives and preference statements:

- For $\alpha, \beta, \gamma \in \mathcal{A}$, we say that $\gamma \doteq \alpha + \beta$ if for all $c \in \mathcal{C}$, $c(\gamma) = c(\alpha) + c(\beta)$.

- For $\varphi, \psi, \chi \in \mathcal{L}^{\mathcal{A}}$, we say that $\varphi \doteq \psi + \chi$ if (i) $\alpha_\varphi \doteq \alpha_\psi + \alpha_\chi$, and $\beta_\varphi \doteq \beta_\psi + \beta_\chi$; and (ii) φ is non-strict if both ψ and χ are non-strict, and otherwise, φ is strict.

5.1 Syntactic deduction $\vdash$ and soundness of inference rules

As usual the proof theory is constructed from axioms and inference rules.

Axioms:

$\alpha \leq \beta$ for all $\alpha, \beta \in \mathcal{A}$ with $\alpha \preceq_{par} \beta$.

Inference rules schemata:

(1) From Strict to Non-Strict: For any $\alpha, \beta \in \mathcal{A}$ the following rule:

From $\alpha < \beta$ deduce $\alpha \leq \beta$.

(2) Addition: For $\chi \in \mathcal{L}^{\mathcal{A}}$ such that $\chi \doteq \varphi + \psi$ the following inference rule

From φ and ψ deduce χ.

(3) Pointwise Multiplication: For any $f \in F$ and $\varphi \in \mathcal{L}^{\mathcal{A}}$ such that $\varphi \doteq f\psi$ the following rule

From ψ deduce φ.

(4) Inconsistent Statement: For any $\alpha \in \mathcal{A}$ and any $\varphi \in \mathcal{L}^{\mathcal{A}}$,

From $\alpha < \alpha$ deduce φ.

(5) Pareto Difference: For any $\psi, \theta \in \mathcal{L}^{\mathcal{A}}$ such that $\psi \preceq_{parD} \theta$:

From ψ deduce θ.

Defining syntactic deduction $\vdash$: Let Γ be a subset of $\mathcal{L}^{\mathcal{A}}$ and $\varphi \in \mathcal{A}$. We say that φ can be proved from Γ, written $\Gamma \vdash \varphi$, if there exists a sequence $\varphi_1, \ldots, \varphi_k$ of elements of $\mathcal{L}^{\mathcal{A}}$ such that $\varphi_k = \varphi$ and for all $i = 1, \ldots, k$, either $\varphi_i \in \Gamma$ or φ_i is an axiom, or there exists an instance of one of the inference rules with consequent φ_i and such that the antecedents are in $\{\varphi_1, \ldots, \varphi_{i-1}\}$. Relation $\vdash$ depends strongly on the set of alternatives $\mathcal{A}$; e.g., $\{\varphi, \psi\} \vdash \varphi + \psi$ (if and) only if $\varphi + \psi \in \mathcal{L}^{\mathcal{A}}$, i.e., only if $\alpha_\varphi + \alpha_\psi$ and $\beta_\varphi + \beta_\psi$ are in $\mathcal{A}$. We write $\vdash$ as $\vdash_{\mathcal{A}}$ if we want to emphasise this dependency. It can happen that for $\Gamma \cup \{\varphi\} \subseteq \mathcal{L}^{\mathcal{A}} \subseteq \mathcal{L}^{\mathcal{B}}$, we have $\Gamma \vdash_{\mathcal{B}} \varphi$, but $\Gamma \nvdash_{\mathcal{A}} \varphi$. (We could also write $\models_{\mathcal{A}}$ to emphasise the dependency on $\mathcal{A}$; however, it isn't usually important to do so, since for $\Gamma \cup \{\varphi\} \subseteq \mathcal{L}^{\mathcal{A}} \subseteq \mathcal{L}^{\mathcal{B}}$, we have $\Gamma \models_{\mathcal{B}} \varphi \iff \Gamma \models_{\mathcal{A}} \varphi$.)

Any given set of alternatives may not be closed under addition (for instance), and there may be $\alpha, \beta \in \mathcal{A}$ with no $\gamma \in \mathcal{A}$ such that $\gamma \doteq \alpha + \beta$. We assume that we can augment $\mathcal{A}$ with additional alternatives, and for any function $g : \mathcal{C} \to \mathbb{Q}^+$, we can construct an alternative α with, for all $c \in \mathcal{C}$, $c(\alpha) = g(c)$.

Next we state a lemma showing soundness of the axioms and inference rules, which is used to prove soundness of the associated syntactic deduction (Proposition 17).

Lemma 16. *Consider any $H \in \mathcal{C}(1)$, any $\alpha, \beta \in \mathcal{A}$, and any $\varphi, \psi, \chi, \theta \in \mathcal{L}^{\mathcal{A}}$.*

(i) If $\alpha \preceq_{par} \beta$ then $H \models \alpha \leq \beta$.

(ii) If $H \models \alpha < \beta$ then $H \models \alpha \leq \beta$.

(iii) If $\chi \doteq \varphi + \psi$, and $H \models \varphi$ and $H \models \psi$ then $H \models \chi$.

(iv) If $\varphi \doteq f\psi$ then $H \models \varphi \iff H \models \psi$.

(v) $H \not\models \alpha < \alpha$.

(vi) If $H \models \psi$ and $\psi \preceq_{parD} \theta$ then $H \models \theta$.

Proof: Write H as $(c_1, \ldots, c_k)$. For $\varphi \in \mathcal{L}^{\mathcal{A}}$ we define i^{φ} to be $k+1$ if for all $i = 1, \ldots, k$, $c_i(\alpha_{\varphi}) = c_i(\beta_{\varphi})$; otherwise, we define i^{φ} to be the minimum i such that $c_i(\alpha_{\varphi}) \neq c_i(\beta_{\varphi})$. Then $\alpha_{\varphi} \equiv_H \beta_{\varphi} \iff i^{\varphi} = k+1$, and $H \models \alpha_{\varphi} < \beta_{\varphi} \iff i^{\varphi} \leq k$ and $c_{i^{\varphi}}(\alpha_{\varphi}) < c_{i^{\varphi}}(\beta_{\varphi})$.

(i): Assume that $\alpha \preceq_{par} \beta$, so that for all $c \in \mathcal{C}$, we have $c(\alpha) \leq c(\beta)$. This implies $\alpha \preceq_H \beta$ and thus $H \models \alpha \leq \beta$.

(ii): Assume that $H \models \alpha < \beta$, so that $\alpha \prec_H \beta$. This implies $\alpha \preceq_H \beta$ and hence $H \models \alpha \leq \beta$.

(iii): Assume that $\chi \doteq \varphi + \psi$, and $H \models \varphi$ and $H \models \psi$.

Case (I): $i^{\varphi} = i^{\psi} = k+1$. Then for all $i = 1, \ldots, k$, $c_i(\alpha_{\varphi}) = c_i(\beta_{\varphi})$ and $c_i(\alpha_{\psi}) = c_i(\beta_{\psi})$. Then, $c_i(\alpha_{\chi}) = c_i(\alpha_{\varphi}) + c_i(\alpha_{\psi}) = c_i(\beta_{\varphi}) + c_i(\beta_{\psi}) = c_i(\beta_{\chi})$, so $i^{\chi} = k+1$, which implies that $\alpha_{\chi} \equiv_H \beta_{\chi}$. We have $\alpha_{\varphi} \equiv_H \beta_{\varphi}$, and also $H \models \varphi$, so φ is non-strict. Similarly, ψ is non-strict. Thus χ is non-strict, and so $H \models \chi$.

Case (II): $i^{\varphi} = i^{\psi} \leq k$. Because $c_{i^{\varphi}}(\alpha_{\varphi}) \neq c_{i^{\varphi}}(\beta_{\varphi})$ and $H \models \varphi$, we have $c_{i^{\varphi}}(\alpha_{\varphi}) < c_{i^{\varphi}}(\beta_{\varphi})$. The same argument implies that $c_{i^{\varphi}}(\alpha_{\psi}) < c_{i^{\varphi}}(\beta_{\psi})$. We then have $c_{i^{\varphi}}(\alpha_{\chi}) < c_{i^{\varphi}}(\beta_{\chi})$, and $i^{\chi} = i^{\varphi}$. This implies that $H \models \alpha_{\chi} < \beta_{\chi}$, and thus, $H \models \chi$, whether χ is strict or non-strict.

Case (III): $i^{\varphi} < i^{\psi}$. Arguing as in Case (II), we have $c_{i^{\varphi}}(\alpha_{\varphi}) < c_{i^{\varphi}}(\beta_{\varphi})$. We also have $c_{i^{\varphi}}(\alpha_{\psi}) = c_{i^{\varphi}}(\beta_{\psi})$. We then have $c_{i^{\varphi}}(\alpha_{\chi}) < c_{i^{\varphi}}(\beta_{\chi})$, and $i^{\chi} = i^{\varphi}$. Again we have $H \models \chi$, whether χ is strict or non-strict.

Case (IV): $i^{\varphi} > i^{\psi}$. This is similar to Case (III), but with the roles of φ and ψ reversed.

(iv): Assume that $\varphi \doteq f\psi$, and consider any $c \in \mathcal{C}$. Because $f(c) > 0$, we have $c(\alpha_{\varphi}) = c(\beta_{\varphi})$ if and only if $c(\alpha_{\psi}) = c(\beta_{\psi})$; and $c(\alpha_{\varphi}) < c(\beta_{\varphi})$ if and only if $c(\alpha_{\psi}) < c(\beta_{\psi})$. This shows that $H \models \varphi \iff H \models \psi$.

(v): $H \not\models \alpha < \alpha$ follows since $\alpha \equiv_H \alpha$ and so $\alpha \not\prec_H \alpha$.

(vi): Suppose that $H \models \psi$ and $\psi \preceq_{parD} \theta$, so that ψ and θ are either both strict or both non-strict; and for all $c \in \mathcal{C}$, $c(\beta_\psi) - c(\alpha_\psi) \leq c(\beta_\theta) - c(\alpha_\theta)$. If it were the case that $i^\psi < i^\theta$ then, because $H \models \psi$, we would have that $c_{i^\psi}(\alpha_\psi) < c_{i^\psi}(\beta_\psi)$ and $c_{i^\psi}(\alpha_\theta) = c_{i^\psi}(\beta_\theta)$, and thus, $c_{i^\psi}(\beta_\psi) - c_{i^\psi}(\alpha_\psi) > 0 = c_{i^\psi}(\beta_\theta) - c_{i^\psi}(\alpha_\theta)$, which contradicts $\psi \preceq_{parD} \theta$. Thus we must have that $i^\psi \geq i^\theta$.

First consider the case when $i^\theta = k + 1$. Then $i^\psi = k + 1$, and so $\alpha_\theta \equiv_H \beta_\theta$ and $\alpha_\psi \equiv_H \beta_\psi$. The latter implies that ψ is non-strict, since $H \models \psi$. Then θ is non-strict and thus, $H \models \theta$.

Now consider the case when $i^\theta \leq k$, and thus $c_{i^\theta}(\alpha_\theta) \neq c_{i^\theta}(\beta_\theta)$. We showed earlier that $i^\theta \leq i^\psi$. If $i^\theta = i^\psi$ then $H \models \psi$ implies that $c_{i^\theta}(\alpha_\psi) < c_{i^\theta}(\beta_\psi)$. If $i^\theta < i^\psi$ then $c_{i^\theta}(\alpha_\psi) = c_{i^\theta}(\beta_\psi)$. So, in either case we have $c_{i^\theta}(\alpha_\psi) \leq c_{i^\theta}(\beta_\psi)$, i.e., $c_{i^\theta}(\beta_\psi) - c_{i^\theta}(\alpha_\psi) \geq 0$. The assumption $\psi \preceq_{parD} \theta$ then implies that $c_{i^\theta}(\beta_\theta) - c_{i^\theta}(\alpha_\theta) \geq 0$, and so, $c_{i^\theta}(\alpha_\theta) \leq c_{i^\theta}(\beta_\theta)$. Since $i^\theta \leq k$ we have $c_{i^\theta}(\alpha_\theta) < c_{i^\theta}(\beta_\theta)$, showing that $H \models \alpha_\theta < \beta_\theta$, and therefore $H \models \theta$ whether θ is strict or non-strict. $\square$

We are now ready to state and prove the soundness result.

Proposition 17. *For $\Gamma \cup \{\varphi\} \subseteq \mathcal{L}^\mathcal{A}$, and any $\mathcal{B} \supseteq \mathcal{A}$, if $\Gamma \vdash_\mathcal{B} \varphi$ then $\Gamma \models_\mathcal{A} \varphi$.*

Proof: First note that if Γ is $\mathcal{C}(1)$-inconsistent, then there is nothing to prove, since $\Gamma \models_\mathcal{A} \varphi$ follows trivially. So, let us assume now that Γ is $\mathcal{C}(1)$-consistent. We use an inductive proof based on Lemma 16. Suppose that $\Gamma \vdash_\mathcal{B} \varphi$. Consider any $H \in \mathcal{C}(1)$ such that $H \models \Gamma$. We need to show that $H \models \varphi$. Since $\Gamma \vdash_\mathcal{B} \varphi$ there exists a sequence $\varphi_1, \ldots, \varphi_k$ of elements of $\mathcal{L}^\mathcal{B}$ such that $\varphi_k = \varphi$ and for all $i = 1, \ldots, k$, either $\varphi_i \in \Gamma$ or φ_i is an axiom, or there exists an instance of one of the inference rules with consequent φ_i and such that the antecedents are in $\{\varphi_1, \ldots, \varphi_{i-1}\}$. Consider any $i \in \{1, \ldots, k\}$. We will prove that, if for all $j < i$, $H \models \varphi_j$ then $H \models \varphi_i$. This then implies that for all $i = 1, \ldots, k$, we have $H \models \varphi_i$, and thus $H \models \varphi_k$, as required.

Therefore, let i be some arbitrary element in $\{1, \ldots, k\}$, and assume that for all $j < i$, $H \models \varphi_j$. We will prove that $H \models \varphi_i$. Let us abbreviate φ_i to be θ. One of the cases (1)–(7) below applies. We consider each case in turn.

(1): θ equals $\alpha \leq \beta$ for some $\alpha, \beta \in \mathcal{B}$, and there exists some $j < i$ with φ_j equalling $\alpha < \beta$. Since $H \models \varphi_j$, by Lemma 16(ii), we have $H \models \alpha \leq \beta$, i.e., $H \models \theta$.

(2): θ equals χ for some $\chi \in \mathcal{L}^\mathcal{B}$ such that $\chi \doteq \varphi + \psi$, and for some $j, l < i$ we have $\varphi = \varphi_j$ and $\psi = \varphi_l$. Since $H \models \varphi_j, \varphi_k$, Lemma 16(iii) implies that $H \models \theta$.

(3): There exists $j < i$ and $f \in F$ such that $\theta \doteq f\varphi_j$. Lemma 16(iv) implies that $H \models \theta$.

(4): There exists $\alpha \in \mathcal{B}$ and $j < i$ such that φ_j equals $\alpha < \alpha$, so we have $H \models \alpha < \alpha$. However, by Lemma 16(v), this is impossible, so Case (4) cannot arise.

(5): There exists $j < i$ such that $\psi = \varphi_j \in \mathcal{L}^{\mathcal{B}}$ and $\psi \preceq_{parD} \theta$. Lemma 16(vi) implies $H \models \theta$.

(6): $\theta \in \Gamma$. Then $H \models \theta$.

(7): θ is equal to $\alpha \leq \beta$ for some $\alpha, \beta \in \mathcal{B}$ such that $\alpha \preceq_{par} \beta$. Lemma 16(i) implies $H \models \theta$.

$\square$

5.2 Completeness of proof theory

We now give a pair of technical lemmas which we will use in the completeness proof.

Lemma 17. *Consider any $\mathcal{C}(1)$-inconsistent $\Gamma \subseteq \mathcal{L}^{\mathcal{A}}$, and suppose that $(\{\varphi_1, \ldots, \varphi_k\}, \mathcal{C}')$ is an inconsistency base for $(\Gamma, \mathcal{C})$, with $\{\varphi_1, \ldots, \varphi_k\}$ being inconsistent. Then there exist strictly positive functions $f_1, \ldots, f_k \in \mathcal{F}$, set of alternatives $\mathcal{B} \supseteq \mathcal{A}$ with $\mathcal{B} - \mathcal{A}$ finite, preference statement $\rho \in \mathcal{L}^{\mathcal{B}}$ and strict preference statement ψ in $\mathcal{L}^{\mathcal{B}}$ such that $\rho \doteq f_1\varphi_1 + \cdots + f_{k-1}\varphi_{k-1}$ and $\psi \doteq f_1\varphi_1 + \cdots + f_k\varphi_k$, and $\Gamma \vdash_{\mathcal{B}} \rho$ and $\Gamma \vdash_{\mathcal{B}} \psi$, and $\beta_\psi \preceq_{par} \alpha_\psi$.*

Proof: Let $T = \{|c(\alpha_{\varphi_i}) - c(\beta_{\varphi_i})| : c \in \mathcal{C}, i \in \{1, \ldots, k\}\} - \{0\}$. If $T = \emptyset$ then set $a = b = 1$, and if $T \neq \emptyset$ let $a = \min T$ and let $b = \max T$, so $0 < a \leq b$. For $i = 1, \ldots, k$ and $c \in \mathcal{C}$, we define $f_i(c) = 1$ if $c(\alpha_{\varphi_i}) > c(\beta_{\varphi_i})$, and otherwise, we define $f_i(c) = d$ where $d = a/(kb) > 0$.

For $i = 1, \ldots, k$ we include elements $\gamma_i, \delta_i, \epsilon_i, \lambda_i$ in $\mathcal{B}$, where $\gamma_i \doteq f_i \alpha_{\varphi_i}$, and $\delta_i \doteq f_i \beta_{\varphi_i}$; and we let $\epsilon_1 \doteq \gamma_1$ and $\lambda_1 \doteq \delta_1$, and for $i = 2, \ldots, k$, $\epsilon_i \doteq \epsilon_{i-1} + \gamma_i$, and $\lambda_i \doteq \lambda_{i-1} + \delta_i$.

There exists $\psi_1 \in \mathcal{L}^{\mathcal{B}}$ with $\psi_1 \doteq f_1\varphi_1$, and $\alpha_{\psi_1} = \gamma_1 = \epsilon_1$ and $\beta_{\psi_1} = \delta_1 = \lambda_1$. Similarly, for $i = 2, \ldots, k$, there exists $\psi_i \in \mathcal{L}^{\mathcal{B}}$ with $\psi_i \doteq \psi_{i-1} + f_i\varphi_i$, and $\alpha_{\psi_i} = \epsilon_i$ and $\beta_{\psi_i} = \lambda_i$.

By the Addition and Pointwise Multiplication rules, for each $i = 1, \ldots, k$, we have $\Gamma \vdash_{\mathcal{B}} \psi_i$. Abbreviate ψ_k to ψ and ψ_{k-1} to ρ. We have $\Gamma \vdash_{\mathcal{B}} \psi$ and $\psi \doteq f_1\varphi_1 + \cdots + f_k\varphi_k$, and $\Gamma \vdash_{\mathcal{B}} \rho$ and $\rho \doteq f_1\varphi_1 + \cdots + f_{k-1}\varphi_{k-1}$. Since $\{\varphi_1, \ldots, \varphi_k\}$ is inconsistent, some φ_i is strict (else the empty model satisfies them all), and therefore, ψ is a strict preference statement.

Consider any $c \in \mathcal{C} - \mathcal{C}'$. By Definition 1(i), $c(\alpha_{\varphi_i}) = c(\beta_{\varphi_i})$ for all $i = 1, \ldots, k$. Thus $c(\alpha_\psi) = c(\beta_\psi)$.

Now consider any $c \in \mathcal{C}'$. For any $j \in \{1, \ldots, k\}$, $c(\alpha_{\varphi_j}) - c(\beta_{\varphi_j}) \geq -b$, and so $c(\gamma_j) - c(\delta_j) \geq -bd = -a/k$. By Definition 1(ii), there exists some $i \in \{1 \ldots, k\}$ such that $c(\alpha_{\varphi_i}) > c(\beta_{\varphi_i})$. This implies that $T \neq \emptyset$. We have $c(\alpha_{\varphi_i}) - c(\beta_{\varphi_i}) \geq a$, and thus

$c(\gamma_i) - c(\delta_i) \geq a > 0$. Now, $c(\alpha_\psi) = \sum_{j=1}^{k} c(\gamma_j)$ and $c(\beta_\psi) = \sum_{j=1}^{k} c(\delta_j)$. Therefore, $c(\alpha_\psi) - c(\beta_\psi) \geq a - (k-1)a/k > 0$. We have shown that for all $c \in \mathcal{C}$, $c(\alpha_\psi) \geq c(\beta_\psi)$, so $\beta_\psi \preceq_{par} \alpha_\psi$. □

Lemma 18. *Suppose $\Gamma \cup \{\varphi\} \subseteq \mathcal{L}^\mathcal{A}$, and that Γ is $\mathcal{C}(1)$-consistent and $\Gamma \models \varphi$. Then there exists $\mathcal{B} \supseteq \mathcal{A}$ (with $\mathcal{B} - \mathcal{A}$ finite), and $\chi, \theta \in \mathcal{L}^\mathcal{B}$ such that $\Gamma \vdash_\mathcal{B} \chi$, and θ is strict and $\theta \doteq \chi + \neg\varphi$, and $\beta_\theta \preceq_{par} \alpha_\theta$.*

Proof: By Lemma 1, $\Gamma \cup \{\neg\varphi\}$ is $\mathcal{C}(1)$-inconsistent. By Corollary 1 there exists an inconsistency base $(\Delta, \mathcal{C}')$ for $(\Gamma \cup \{\neg\varphi\}, \mathcal{C})$ with Δ being a finite and $\mathcal{C}(1)$-inconsistent subset of $\Gamma \cup \{\neg\varphi\}$, and $\mathcal{C}' \subseteq \mathcal{C}$. Now, Δ contains $\neg\varphi$, since Δ is $\mathcal{C}(1)$-inconsistent and Γ is $\mathcal{C}(1)$-consistent. We write Δ as $\{\varphi_1, \ldots, \varphi_k\}$ with $\varphi_k = \neg\varphi$.

By Lemma 17, there exist strictly positive functions $f_1, \ldots, f_k \in F$, set of alternatives $\mathcal{B} \supseteq \mathcal{A}$ with $\mathcal{B} - \mathcal{A}$ finite, preference statement $\rho \in \mathcal{L}^\mathcal{B}$ and strict preference statement ψ in $\mathcal{L}^\mathcal{B}$ such that $\rho \doteq f_1 \varphi_1 + \cdots + f_{k-1} \varphi_{k-1}$ and $\psi \doteq f_1 \varphi_1 + \cdots + f_k \varphi_k$, $\Gamma \vdash_\mathcal{B} \rho$ and $\Gamma \vdash_\mathcal{B} \psi$, and $\beta_\psi \preceq_{par} \alpha_\psi$.

Let $\mathcal{B}' = \mathcal{B} \cup \{\alpha_\chi, \beta_\chi, \alpha_\theta, \beta_\theta\}$, where $\alpha_\chi \doteq \frac{1}{f_k} \alpha_\rho$ and $\beta_\chi \doteq \frac{1}{f_k} \beta_\rho$, and $\alpha_\theta \doteq \alpha_\chi + \beta_\varphi$ and $\beta_\theta \doteq \beta_\chi + \alpha_\varphi$, and χ, θ (which are thus in $\mathcal{L}^{\mathcal{B}'}$) are such that $\chi \doteq \frac{1}{f_k} \rho$ and $\theta \doteq \chi + \neg\varphi$, i.e., $\theta \doteq \chi + \varphi_k$. We have $f_k \theta \doteq f_k \chi + f_k \varphi_k \doteq \rho + f_k \varphi_k$ and thus $\psi \doteq f_k \theta$. This implies that θ is a strict statement and that $\beta_\theta \preceq_{par} \alpha_\theta$. Now, $\Gamma \vdash_\mathcal{B} \rho$ implies that $\Gamma \vdash_{\mathcal{B}'} \rho$ (because $\mathcal{B}' \subseteq \mathcal{B}$). Since $\chi \doteq \frac{1}{f_k} \rho$, we have $\Gamma \vdash_{\mathcal{B}'} \chi$, using the Pointwise Multiplication inference rule, completing the proof. □

These lemmas lead to the completeness theorems.

Theorem 4. *Consider any $\Gamma \subseteq \mathcal{L}^\mathcal{A}$ and any $\varphi \in \mathcal{L}^\mathcal{A}$. Then there exists $\mathcal{B} \supseteq \mathcal{A}$, with $\mathcal{B} - \mathcal{A}$ finite such that $\Gamma \models \varphi \iff \Gamma \vdash_\mathcal{B} \varphi$.*

Proof: $\Leftarrow$ follows by Proposition 17. To prove the converse, let us assume that $\Gamma \models \varphi$; we will show that $\mathcal{A}$ can be extended to $\mathcal{B}$ such that $\Gamma \vdash_\mathcal{B} \varphi$.

First let us consider the case when Γ is $\mathcal{C}(1)$-inconsistent. By Corollary 1 there exists $\mathcal{C}' \subseteq \mathcal{C}$ and a $\mathcal{C}(1)$-inconsistent subset $\{\varphi_1, \ldots, \varphi_k\}$ of Γ, such that $(\{\varphi_1, \ldots, \varphi_k\}, \mathcal{C}')$ is an inconsistency base for $(\Gamma, \mathcal{C})$. By Lemma 17, there exist strictly positive functions $f_1, \ldots, f_k \in F$, set of alternatives $\mathcal{B} \supseteq \mathcal{A}$ with $\mathcal{B} - \mathcal{A}$ finite, and strict preference statement ψ in $\mathcal{B}$ such that $\psi \doteq f_1 \varphi_1 + \cdots + f_k \varphi_k$, and $\Gamma \vdash_\mathcal{B} \psi$ and $\beta_\psi \preceq_{par} \alpha_\psi$. Consider any $\gamma \in \mathcal{A}$. Then $\beta_\psi \preceq_{par} \alpha_\psi$ implies for all $c \in \mathcal{C}$, $c(\beta_\psi) - c(\alpha_\psi) \leq 0 = c(\gamma) - c(\gamma)$. The Pareto Difference inference rule then implies that $\Gamma \vdash_\mathcal{B} \gamma < \gamma$, since ψ is strict, and hence, by the Inconsistent Statement inference rule, $\Gamma \vdash_\mathcal{B} \varphi$, as required.

Now we consider the case when Γ is $\mathcal{C}(1)$-consistent. By Lemma 18, we have that there exists set of alternatives $\mathcal{B} \supseteq \mathcal{A}$ with $\mathcal{B} - \mathcal{A}$ finite, and $\chi, \theta \in \mathcal{L}^\mathcal{B}$ such that $\Gamma \vdash_\mathcal{B} \chi$, and θ is strict, $\theta \doteq \chi + \neg\varphi$, and $\beta_\theta \preceq_{par} \alpha_\theta$. Then, by definition of $\neg\varphi$, we have $\alpha_\theta \doteq \alpha_\chi + \beta_\varphi$ and $\beta_\theta \doteq \beta_\chi + \alpha_\varphi$. This implies that for all $c \in \mathcal{C}$, $c(\beta_\chi) + c(\alpha_\varphi) \leq c(\alpha_\chi) + c(\beta_\varphi)$, and thus, for all $c \in \mathcal{C}$, $c(\beta_\chi) - c(\alpha_\chi) \leq c(\beta_\varphi) - c(\alpha_\varphi)$. Now, since $\theta \doteq \chi + \neg\varphi$ and θ is strict, if χ is non-strict then $\neg\varphi$ must be strict and so φ is non-strict. The Pareto Difference inference rule then implies that $\Gamma \vdash_\mathcal{B} \varphi$. If, on the other hand, χ is strict then the Pareto Difference inference rule implies that $\Gamma \vdash_\mathcal{B} \alpha_\varphi < \beta_\varphi$, and thus $\Gamma \vdash_\mathcal{B} \alpha_\varphi \leq \beta_\varphi$, using the From Strict to Non-Strict rule. Therefore $\Gamma \vdash_\mathcal{B} \varphi$ whether φ is strict or non-strict. □

Let $\mathcal{A}^*$ be a set of alternatives including for each function $g : \mathcal{C} \to \mathbb{Q}^+$, an alternative α with, for all $c \in \mathcal{C}$, $c(\alpha) = g(c)$, and let $\mathcal{A}' = \mathcal{A} \cup \mathcal{A}^*$. Consider any $\Gamma \subseteq \mathcal{L}^\mathcal{A}$ and any $\varphi \in \mathcal{L}^\mathcal{A}$. Then $\Gamma \cup \{\varphi\} \subseteq \mathcal{L}^{\mathcal{A}'}$. If we use $\mathcal{A}'$ instead of $\mathcal{A}$ in the proofs of Lemma 17 and 18, and Theorem 4, we can use $\mathcal{B} = \mathcal{A}'$ in each case. This leads, for arbitrary Γ and φ, to: $\Gamma \models_{\mathcal{A}'} \varphi \iff \Gamma \vdash_{\mathcal{A}'} \varphi$, which since $\Gamma \models_{\mathcal{A}'} \varphi$ holds if and only if $\Gamma \models_\mathcal{A} \varphi$ holds, gives the following version of the completeness result.

Theorem 5. *For any $\mathcal{A}$, there exists $\mathcal{A}' \supseteq \mathcal{A}$ such that for any $\Gamma \subseteq \mathcal{L}^\mathcal{A}$ and any $\varphi \in \mathcal{L}^\mathcal{A}$, $\Gamma \models \varphi \iff \Gamma \vdash_{\mathcal{A}'} \varphi$.*

Discussion of related preference inference based on weighted sum

Another natural notion of preference inference, which is similar to that defined e.g., in [13, 12], is based on weighted sums. In each model a non-negative weight is assigned to each evaluation, and the overall desirability of an alternative is the weighted sum of the evaluations on the alternative. More precisely, let the set of models be the set of functions e from $\mathcal{C}$ to $\mathbb{Q}^+$. We say that e *satisfies* $\alpha \leq \beta$ if $\sum_{c \in \mathcal{C}} e(c)c(\alpha) \leq \sum_{c \in \mathcal{C}} e(c)c(\beta)$. Similarly, we say that e *satisfies* $\alpha < \beta$ if $\sum_{c \in \mathcal{C}} e(c)c(\alpha) < \sum_{c \in \mathcal{C}} e(c)c(\beta)$. As for the other kinds of preference inference, we say, for $\Gamma \cup \{\varphi\} \subseteq \mathcal{L}^\mathcal{A}$, that Γ entails φ if e satisfies φ for every e satisfying Γ. This preference inference satisfies the above axiom schema, and all the inference rules except for (3) Pointwise Multiplication (and thus is weaker than $\models_{\mathcal{C}(1)}$). Instead a weaker form of (3) holds, based on using only constant functions f. The Pointwise Multiplication inference rule might thus be considered as characteristic of preference inference based on simple lexicographic models.

6 Discussion and conclusions

We defined a class of relatively simple preference logics based on hierarchical models. These generate an adventurous form of inference, which can be helpful if there is only relatively sparse input preference information. We showed that the complexity of preference deduction is coNP-complete in general, and polynomial for the case where the criteria are assumed to be totally ordered (the simple lexicographic models case, Section 4).

The latter logic has strong connections with the preference inference formalism described in [18]. To clarify the connection, for each evaluation $c \in \mathcal{C}$ we can generate a variable X_c, and let V be the set of these variables. For each alternative $\alpha \in \mathcal{A}$ we generate a complete assignment α^* on the variables V (i.e., an outcome as defined in [18]) by $\alpha^*(X_c) = c(\alpha)$ for each $X_c \in V$. Note that values of $\alpha^*(X_c)$ are non-negative numbers, and thus have a fixed ordering, with zero being the best value. A preference statement $\alpha \leq \beta$ in $\mathcal{L}_{\leq}^{\mathcal{A}}$ then corresponds with a basic preference formula $\alpha^* \geq \beta^*$ in [18]. Each model $H \in \mathcal{C}(1)$ corresponds to a sequence of evaluations, and thus has an associated sequence of variables; this sequence together with the fixed value orderings, generates a lexicographic model as defined in [18].

In contrast with the lexicographic inference system in [18], the logic developed in this paper allows strict (as well as non-strict) preference statements, and allows more than one variable at the same level. However, the lexicographic inference logic from [18] does not assume a fixed value ordering (which, translated into the current formalism, corresponds to not assuming that the values of the evaluation function are known); it also allows a richer language of preference statements, where a statement can be a compact representation for a (possibly exponentially large) set of basic preference statements of the form $\alpha \leq \beta$. Many of the results of Section 4 immediately extend to richer preference languages (by replacing a preference statement by a corresponding set of basic preference statements). In future work we will determine under what circumstances deduction remains polynomial when extending the language, and when removing the assumption that the evaluation functions are known.

The coNP-hardness result for the general case (and for the $\models_{\mathcal{C}(t)}^{\oplus}$ systems with $t \geq 2$) is notable and perhaps surprising, since these preference logics are relatively simple ones. The result obviously extends to more general systems. The preference inference system described in [16] is based on much more complex forms of lexicographic models, allowing conditional dependencies, as well as having local orderings on sets of variables (with bounded cardinality). Theorem 1 implies that the (polynomial) deduction system in [16] is not more general than the system described here (assuming $P \neq NP$). It also implies that if one were to extend the system from [16] to allow a richer form of equivalence, generalising e.g., the $\models_{\mathcal{C}(2)}^{\oplus}$ system, then the preference inference will no longer be polynomial.

Acknowledgements

This publication has emanated from research conducted with the financial support of Science Foundation Ireland (SFI) under Grant Number SFI/12/RC/2289. Nic Wilson was also supported by the School of EEE&CS, Queen's University Belfast. We are grateful to the reviewers for their helpful comments.

References

[1] M. Bienvenu, J. Lang, and N. Wilson. From preference logics to preference languages, and back. In *Proc. KR 2010*, 2010.

[2] R. Booth, Y. Chevaleyre, J. Lang, J. Mengin, and C. Sombattheera. Learning conditionally lexicographic preference relations. In *Proc. ECAI-2010*, pages 269–274, 2010.

[3] C. Boutilier, R. I. Brafman, C. Domshlak, H. Hoos, and D. Poole. CP-nets: A tool for reasoning with conditional *ceteris paribus* preference statements. *Journal of Artificial Intelligence Research*, 21:135–191, 2004.

[4] S. Bouveret, U. Endriss, and J.Lang. Conditional importance networks: A graphical language for representing ordinal, monotonic preferences over sets of goods. In *Proc. IJCAI-09*, pages 67–72, 2009.

[5] R. Brafman, C. Domshlak, and E. Shimony. On graphical modeling of preference and importance. *Journal of Artificial Intelligence Research*, 25:389–424, 2006.

[6] M. Bräuning and E. Hüllermeier. Learning conditional lexicographic preference trees. In *Preference Learning (PL-12), ECAI-12 workshop*, 2012.

[7] D. Bridge and F. Ricci. Supporting product selection with query editing recommendations. In *RecSys '07*, pages 65–72, New York, NY, USA, 2007. ACM.

[8] J. Dombi, C. Imreh, and N. Vincze. Learning lexicographic orders. *European Journal of Operational Research*, 183(2):748–756, 2007.

[9] J. Figueira, S. Greco, and M. Ehrgott. *Multiple Criteria Decision Analysis—State of the Art Surveys*. Springer International Series in Operations Research and Management Science Volume 76, 2005.

[10] P. A. Flach and E.T. Matsubara. A simple lexicographic ranker and probability estimator. In *Proc. ECML-2007*, pages 575–582, 2007.

[11] J. Fürnkranz and E. Hüllermeier (eds.). *Preference Learning*. Springer-Verlag, 2010.

[12] Anne-Marie George, Abdul Razak, and Nic Wilson. The comparison of multi-objective preference inference based on lexicographic and weighted average models. In *27th IEEE International Conference on Tools with Artificial Intelligence, ICTAI 2015, Vietri sul Mare, Italy, November 9-11, 2015*, pages 88–95, 2015.

[13] R. Marinescu, A. Razak, and N. Wilson. Multi-objective constraint optimization with tradeoffs. In *Proc. CP-2013*, pages 497–512, 2013.

[14] W. Trabelsi, N. Wilson, D. Bridge, and F. Ricci. Preference dominance reasoning for conversational recommender systems: a comparison between a comparative preferences and a sum of weights approach. *International Journal on Artificial Intelligence Tools*, 20(4):591–616, 2011.

[15] Molly Wilson and Alan Borning. Hierarchical constraint logic programming. *The Journal of Logic Programming*, 16(3-4):277–318, 1993.

[16] N. Wilson. Efficient inference for expressive comparative preference languages. In *Proc. IJCAI-09*, pages 961–966, 2009.

[17] N. Wilson. Computational techniques for a simple theory of conditional preferences. *Artificial Intelligence*, 175(7-8):1053–1091, 2011.

[18] N. Wilson. Preference inference based on lexicographic models. In *Proc. ECAI-2014*, pages 921–926, 2014.

[19] N. Wilson, A.-M. George, and B. O'Sullivan. Computation and complexity of preference inference based on hierarchical models. In *Proc. IJCAI-2015*, 2015.

A Semantic Approach to Combining Preference Formalisms

Alireza Ensan
Simon Fraser University, Canada
aensan@sfu.ca

Eugenia Ternovska
Simon Fraser University, Canada
ter@sfu.ca

Abstract

We propose a versatile framework for combining knowledge bases in modular systems with preferences. In our formalism, each module (knowledge base) can be specified in a different language. We define the notion of a preference-based modular system that includes a formalization of meta-preferences. We prove that our formalism is robust in the sense that the operations for combining modules preserve the notion of a preference-based modular system. Finally, we formally demonstrate correspondences between our framework and the related preference formalisms of cp-nets, preference-based planning, and answer set optimization. Our framework allows one to use these preference formalisms (and others) in combination, in the same modular system.

1 Introduction

Combining knowledge bases (KBs) is very important when common sense reasoning is involved. For example, in planning, we may want to combine temporal and spatial reasoning, or reasoning from the point of view of several agents. Here, we consider each knowledge base, that is called module, as a set of structures.[1] Modular Systems (MS) [28] is a framework to combine heterogeneous knowledge bases. Modules are combined through the operators of production, union, projection, complementation and selection. An algorithm for finding models of MSs was proposed in [29]. An improvement of the algorithm, in the

---

[1] A structure, e.g. $\mathcal{A} = (A, R_1^{\mathcal{A}}, ..., R_n^{\mathcal{A}}, f_1^{\mathcal{A}}, ..., f_m^{\mathcal{A}}, c_1^{\mathcal{A}}, ..., c_l^{\mathcal{A}})$ is a domain A together with an interpretation function $\mathcal{I}$ of relations (R) such that $R^{\mathcal{A}} = \mathcal{I}(R) \subseteq A^n$, function symbols ($f$) where $f^{\mathcal{A}} = \mathcal{I}(f) : A^m \to A$, and constants ($c$) where $c^{\mathcal{A}} = \mathcal{I}(c) \in A$.

same paper, uses approximations to reduce the search space. Connections to Satisfiability Modulo Theory and other systems were discussed in [29]. In [20] lazy clause generation has been applied to decide if a given modular system has any model.

An important aspect of knowledge representation systems is the capability to represent preferences. The literature presents a variety of approaches to formalize preferences, e.g. [6], [24], [12], [10], [25], [4], [13], [32] and [15]. Several surveys have appeared in recent years categorizing preference formalisms from various perspectives. For example, in [2], a set of preference formalisms for planning have been introduced. The authors of [11] classified preference frameworks in non-monotonic reasoning.

Preferences in database systems have been studied by different researchers such as [18], [3] and [27]. A primary well-known preference language in database systems was proposed in [18]. In this language, some preference constructors were introduced to express basic preference terms. For example, consider the following statement: to buy a car, it is preferred to have BMW or Mercedes-Benz than other models. POS is a constructor that can be used to express this statement. In the aforementioned example, POS includes BMW and Mercedes-Benz. Formally speaking, given two n-arity tuples $A = (a_1, ..., a_n)$, $B = (b_1, ..., b_n)$ and a set called POS, A is preferred to B (notation $A >_P B$) with respect to i^{th} attribute (column) in the database table if and only if $A[i] \in POS$ and $B[i] \notin POS$. Likewise, P is a NEG preference and $A >_P B$ if and only if $A[i] \notin NEG$ and $B[i] \in NEG$. In addition to POS, Pareto and prioritized accumulation are two operators broadly used in several frameworks. Prioritized accumulation that is denoted by & gives priority to a preference. Let A and B be tuples of the same relational schema R. A is preferred to B (notation $A >_{P_1 \& P_2} B$) if and only if $A >_{P_1} B \vee (A \not>_{P_1} B \wedge B \not>_{P_1} A \wedge A <_{P_2} B)$. Pareto operator combines two preferences such that A is preferred to to B with respect to composition of P_1 and P_2 ($P = P_1 \otimes P_2$) if and only if $(A >_{P_1} B \wedge \neg(B >_{P_2} A) \vee (A >_{P_2} B \wedge \neg B >_{P_1} A)$.

In practical settings, systems such as web services, planners, business process controllers, and so on consist of intricate interconnected parts. Each component may interact with other parts and its associated knowledge base may be updated during a process being executed. Current frameworks are incapable of modelling preferences in such dynamic environments. For example, in [8] and [9], preferences are defined by some rules in Answer Set Programming (ASP). However, a knowledge base of ASP preferences and rules, cannot be updated or connected to other systems. Similarly, in [33] a language for preference representation and inference is proposed but the framework does not address how to do reasoning when preferences are combined. Some attempts have been made to combine preferences such as in [23], [18], and [1]. But these frameworks have two main shortcomings: 1) preferences of two components can be combined only when their languages are the same. For instance in [23] and [18] preferences are expressed in first-order logic. So these languages cannot model heterogeneous data systems such as web services that each part may have its own language, 2) two components are sequentially composed. Nonetheless, many real world systems have more complicated structure (see feedback connection in Example 1 below). The following example clarifies the complexity of formalizing a modular system with preferences. In the real word, a complex can interact in much more complicated way, such as feedback in Example 1.

Example 1. *A Logistic Service Provider (LSP) is a modular system that can be used by a company that provides logistic services. It decides how to pack goods and deliver them. It solves two NP-complete tasks interactively, – Multiple Knapsack (module M_K) and Travelling Salesman Problem (module M_{TSP}). The system takes orders from customers (items $Items(i)$ to deliver, their prices $p(i)$, weights $w(i)$), and the capacity of trucks available $c(t)$, decides how to pack $Pack(i,t)$ items in trucks, and for each truck, solves a TSP problem. The feedback about solvability of TSP is sent back to M_K. Module M_{TSP} takes a candidate solution from M_K, together with the graph of cities and routes with distances, allowable distance limit and destinations for each product. The output of this module is the route, for each truck $Route(t, n, c_r)$, where t is a truck, n is the number in the sequence, and c_r is a city. The Knapsack problem is written, in, e.g. Integer Linear Programming (ILP), and TSP in Answer Set Programming (ASP). The modules M_K and M_{TSP} are composed in sequence, with a feedback going from an output of M_{TSP} to an input of M_K. A solution to the compound module, M_{LSP}, to be acceptable, must satisfy both sub-systems. The company may have preferences for packing and delivery of products. E.g. if a fragile item is packed in a truck, it may be preferred to exclude heavy items. Among certain routes with equal costs, some of them may be preferred to others. It is possible that preferences in the Knapsack problem are formalized by cp-nets [4] and the TSP's preferences are represented in preference-based Answer Set Programming framework [9]. In Figure 1, $\mathcal{P}_k$ denotes the preferences of the knapsack module and $\mathcal{P}_{TSP}$ denotes the TSP module's preferences. Formalizing this modular problem with preferences is not easy because: 1) the Knapsack and the TSP are axiomatized in different languages, 2) preferences of each module are represented by a different formalism, 3) preference formalisms use different languages than the axiomatizations of the modules themselves, 4) two modules communicate in a complex way that includes a feedback loop from M_{TSP} to M_K.*

1.1 Contributions

We propose model-theoretic foundations for combining KBs with preferences in modular systems. On the logic level, each module is represented by a KB in some logic $\mathcal{L}$, and its preferences (and meta-preferences) are represented by (strict) partial orders on partial structures in a preference formalism named $\mathcal{P}\text{-}\mathcal{MS}$.[2] Different logics and preference formalisms

---

[2] Any logic with model-theoretic semantics can be used, including logic programs.

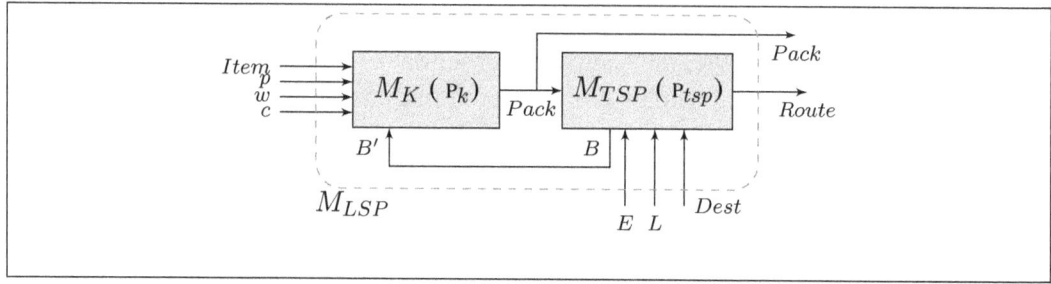

Figure 1: Logistics Service Provider $M_{LSP} = \sigma_{B \equiv B'}(M_K \times M_{TSP})$

can be used for modules in the same system. Operations for combining modules are generalized to combine preferences of each module. We prove that our formalism is *closed* in the sense that the operations for combining modules preserve the notion of a preference-based modular system. Our formalism is consistent with (and extends) the model-theoretic semantics of modular systems [28]. In model theoretic semantics, each module is viewed as a set of structures. We also prove that our formalism represent cp-nets, preference-based planning, and answer set optimization. Thus we can combine them in one modular system.

1.2 Novelty

With our formalism, *each module can be formalized in a different framework*. To our knowledge, this is the first multi-language preference formalism. This generality is achieved through the model-theoretic semantics of modular systems. Another novelty is the ability of handling preferences in complexly structured systems. For instance, in Example 1, there is a complex combination of Knapsack and TSP problems (feedback from TSP to Knapsack). In contrast, these complex systems were not representable in previous work. E.g., in [18].

1.3 Paper Structure

After giving preliminaries, we discuss how preferences in an atomic module can be represented. Then we extend the idea to modular systems and study preference specification in modular systems when some preferences of their components are given. After that, we formally analyze three well-known preference formalisms through our approach to preferences. We conclude the paper with a summary and future work.

This paper significantly extends IJCAI paper [14] by adding complete and detailed proofs and novel examples illustrating how the formalism generalizes and combines other approaches to handling preferences. We also use a newer version of the algebra of modular systems [31, 30] to which our preference framework is adapted.

2 Algebra of Modular Systems

In this section we briefly review some aspects of modular systems proposed in [31], [30], [28], [29], and [21].

A *vocabulary* (denoted, e.g. $\tau, \sigma, \varepsilon$) is a set of non-logical (predicate and function) symbols. A τ-structure is a domain (a set), and interpretation of vocabulary symbols in τ. We use notations $vocab(\mathcal{A})$, $vocab(\phi)$, $vocab(M)$ to denote the vocabulary of structure $\mathcal{A}$, formula ϕ and module M, respectively, and we use $\mathcal{A}|_\sigma$ to mean structure $\mathcal{B}$ restricted to vocabulary σ.

Partial structures allow interpretation of some vocabulary symbols to be partially specified. The idea of partial structures originates from the notion of three-valued logic that a truth value of a statement can be true, false, or unknown [19]. The algorithm for solving modular systems [29] constructs expansions incrementally, by adding information to partial structures.

Definition 1. *$\mathcal{B}$ is a τ_p-partial structure over vocabulary τ if: (1) $\tau_p \subseteq \tau$, (2) $\mathcal{B}$ gives a total interpretation to symbols in $\tau \setminus \tau_p$, and (3) for each n-ary symbol R in τ_p, $\mathcal{B}$ interprets R using two sets R^+ and R^- such that $R^+ \cap R^- = \emptyset$, and $R^+ \cup R^- \neq [dom(\mathcal{B})]^n$.*

For two partial structures $\mathcal{B}$ and $\mathcal{B}'$ over the same vocabulary and domain, we say that $\mathcal{B}$ extends $\mathcal{B}'$ if all undefined symbols in $\mathcal{B}$ are also undefined in $\mathcal{B}'$.

Notation 1. *Let $V = \{a_1, a_2, ..., a_n\}$ be a set of vocabulary symbols. Let $\mathcal{A}$ be a partial structure that interprets a subset $X \subseteq V$ such that V-X is undefined. Each $a_i \in X$ can be interpreted as false, represented by a_i^-, or as true, represented by a_i or a_i^+. Suppose Y is a set of the form $\{a_i^{\mathcal{A}} | a_i \in X\}$ where $a_i^{\mathcal{A}} = a_i^+$ or a_i^- is an interpretation of a_i by $\mathcal{A}$. We assume that Y is the representation of $\mathcal{A}$.*

2.1 Syntax of Modular Systems

In our framework, a module is considered as a class of structures. Let $\tau_M = \{M_1, M_2, ...\}$ be a fixed vocabulary of atomic module symbols and let τ be a fixed vocabulary. Algebraic expression for modules are built by the grammar:

$$E ::= \bot \mid M_i \mid E \times E \mid E + E \mid -E \mid \pi_\delta E \mid \sigma_\Theta E.$$

We call $\times$ product, $+$ union, $-$ complement, π_δ projection onto δ, and $\sigma R \equiv Q$ selection. Modules that are not atomic are called compound. Each atomic module symbol M_i has an associated vocabulary $vocab(M_i) \subseteq \tau$. Vocabulary of a compound module is given by $vocab(\bot) = \tau$, $vocab(E_1 \times E_2) = vocab(e_1) \cap vocab(E_2)$, $vocab(E_1 + E_2) = vocab(e_1) \cup$

$vocab(E_2)$, $vocab(-E) = vocab(E)$, $vocab(\pi_\delta E) = \delta$, and $vocab(\sigma_\Theta E) = vocab(E)$. The three set-theoretic operations are union (+), intersection (×), complementation (-). Projection (π_δ) is a family of unary operations, one for each δ. Each relational symbol (constant or variable) in δ must appear in E. The operation restricts each structure $\mathcal{A}$ of M to $\mathcal{A}|_\delta$ to leaving the interpretation of other symbols open. Thus, it increases the number of models. The condition Θ in selection is an expression of the form $R \equiv Q$, where $R, Q \in vocab(E)$. Selection can model feedback in dynamic systems when output Q is connected to input R. A module can be given by any decision procedure, be a set of models of a KB, be given by an inductive definition, a C or an ASP program, or by an agent making decisions.

2.2 Semantics of Modular Systems

Let $\mathcal{C}$ be the set of all τ-structures with domain A. Modules are interpreted by subsets of $\mathcal{C}$. A module interpretation assigns to each atomic module $M_i \in \tau_M$ a set of τ-structures such that two τ-structures $\mathcal{A}_1$ and $\mathcal{A}_2$ that coincide on $vocab(M_i)$ satisfying $\mathcal{A}_1 \in \mathcal{I}(M)$, where $\mathcal{I}(M)$ denotes the interpretation of M as a set of structures by $\mathcal{I}$, iff $\mathcal{A}_2 \in \mathcal{I}(M)$. The value of a modular expression E in $\mathcal{I}$, denoted $[\![E]\!]^\mathcal{I}$ is defined as follows.

$[\![\perp]\!]^\mathcal{I} = \emptyset$
$[\![M_i]\!]^\mathcal{I} = \mathcal{I}(M_i)$
$[\![E_1 \times E_2]\!]^\mathcal{I} = [\![E_1]\!]^\mathcal{I} \cap [\![E_2]\!]^\mathcal{I}$
$[\![E_1 + E_2]\!]^\mathcal{I} = [\![E_1]\!]^\mathcal{I} \cup [\![E_2]\!]^\mathcal{I}$
$[\![-E]\!]^\mathcal{I} = \mathcal{C} - [\![E]\!]^\mathcal{I}$
$[\![\pi_\delta(E)]\!]^\mathcal{I} = \{\mathcal{A} \mid \exists \mathcal{A}' \ (\mathcal{A}' \in [\![E]\!]^\mathcal{I} \text{ and } \mathcal{A}|_\delta = \mathcal{A}'|_\delta)\}$
$[\![\sigma_{Q \equiv R} E]\!]^\mathcal{I} = \{\mathcal{A} \mid \mathcal{A} \in [\![E]\!]^\mathcal{I} \text{ and } Q^\mathcal{A} = R^\mathcal{A}\}$

We call $\mathcal{A}$ a *model* of E in $\mathcal{I}$ (denoted $\mathcal{A} \models_\mathcal{I} E$) if $\mathcal{A} \in [\![E]\!]^\mathcal{I}$. From now, we assume that a module interpretation $\mathcal{I}$ is given and fixed. Slightly abusing notation, we often omit reference to $\mathcal{I}$ and write, e.g., $\mathcal{A} \models E$ if $\mathcal{A} \in E$ instead of $\mathcal{A} \models_\mathcal{I} E$ when $\mathcal{A} \in [\![E]\!]^\mathcal{I}$.

Model-theoretic semantic associates, with each modular system, a set of structures. We assume that the domains of all structures are included in a (potentially infinite) universal domain U.

2.3 Algebra with Information Flow

Modules that have inputs and outputs are very common. Many software programs and hardware devices are of that form. In the Logistics Service Provider (Example 1), e.g. users' requests could be on the input, and the truck rout and packing solutions on the output. In this section, we add information propagation to the algebra, so that modules become binary higher-order input-output relations. This version of the algebra may be called "dynamic". Fixing an input and an output vocabularies in some modules allows us to talk about model expansion (MX) task [21]. In this task, a given structure, which might have an empty vocabulary, is expanded with interpretations of new vocabulary symbols to satisfy a specification. Complexity-wise, MX lies in-between model checking (full structure is given) and satisfiability (no structure is given). The task generalizes to the formalism of Modular Systems.

Definition 2 (Model Expansion (MX) Task). *Given:* $\mathcal{B}|_\sigma$ *(instance structure) and algebraic expression α with input symbols σ. Find: $\mathcal{B}$ such that $\mathcal{B}$ satisfies α. Structure $\mathcal{B}$ expands structure $\mathcal{B}|_\sigma$ and is called a solution of modular system α for a particular input $\mathcal{B}|_\sigma$.*

Thus, the algebra with information flow may be called "a logic of hybrid MX tasks", and it will be interpreted over transition systems.

Example 2. *In the graph 3-coloring problem (with three possible colors including red, blue, and green), the problem axiomatization α in first order logic is specified as follows:*

$$\alpha = \forall x \, [(R(x) \vee B(x) \vee G(x)) \\ \wedge \neg((R(x) \wedge B(x)) \vee (R(x) \wedge G(x)) \vee (B(x) \wedge G(x)))] \\ \wedge \forall x \forall y \, [E(x,y) \supset (\neg(R(x) \wedge R(y)) \\ \wedge \neg(B(x) \wedge B(y)) \wedge \neg(G(x) \wedge G(y)))].$$

Where R, G, B denote red, green, and blue colors respectively. An instance structure $\mathcal{A} = \mathcal{B}|_\sigma$ for vocabulary $\sigma = \{E\}$ is a graph $\mathcal{G} = (V, E)$. The MX task is to find an interpretation for the symbols $\{R, B, G\}$ satisfying α:

$$\underbrace{(\overbrace{V; E^\mathcal{A}}^{\mathcal{A}}, R^\mathcal{B}, B^\mathcal{B}, G^\mathcal{B})}_{\mathcal{B}} \models \alpha.$$

The structures $\mathcal{B}$ which satisfy α are 3-colourings of $\mathcal{G}$.

Model Expansion tasks are common in AI such as in planning, scheduling, logistics, supply chain management, etc. Java programs, if they are of input-output type, can be viewed as model expansion tasks, regardless of what they do internally. ASP systems, e.g., Clasp [17] mostly solve model expansion, and so do CP languages such as Essence [16], as shown in [22]. CSP in the traditional AI form (respectively, in the homomorphism form) is representable by model expansion where mappings to domain elements (respectively, homomorphism functions) are expansion functions.

3 $\mathcal{P}$-$\mathcal{MS}$: Preference-based Modular Systems

In this section, we define a *primitive* preference $\mathcal{P}$ in an atomic module M. Then, binary relation $\succ_\mathcal{P}$ is introduced to compare structures in M with respect to $\mathcal{P}$. After that, we consider a primitive module M with a set of *primitive* preferences $\Pi = \{\mathcal{P}_1, ..., \mathcal{P}_n\}$ and a preference relation $\mathcal{MP}$ on elements of Π. In this setting, a preferred structure is defined based on binary relation $\succ_{\mathcal{MP}}$. For modular systems, we introduce the notion of *compound* preferences constructed by modular systems operators. A modular system with preferences (*primitive* or *compound*) is called a Preference-based Modular Systems ($\mathcal{P}$-$\mathcal{MS}$). We show that when we combine modular systems with preferences, the result is also a $\mathcal{P}$-$\mathcal{MS}$.

To have a formalism compatible with model theoretic semantics of modular systems, we define preference statements based on the concept of structures. However, using structures to model preferences is not always practical. Formally speaking, some interpreted symbols may be preferred to others, and there could not be enough information to decide about the rest. Unlike structures, partial structures interpret a subset of vocabulary symbols, while interpretation of other symbols is unknown. In our formalism, a preference statement can be represented by a partial order over a set of partial structures when certain conditions hold. First, we explain the meaning of strict partial order.

Definition 3. *A strict partial order $\mathcal{O}$ over a set $\mathcal{S}$ is a pair $\mathcal{O} := (\mathcal{S}, \prec)$ such that $\prec$ is a binary relation over elements of $\mathcal{S}$ that is anti-reflexive, asymmetric and transitive.*

Preference statements in natural language are represented as conditional statements, e.g., *Mary prefers to buy Ford to Toyota if car's body-type is SUV or coupe. If it is sedan, Mary prefers Toyota to Ford.* We present this statement as $\mathcal{P} = (\mathcal{O}, \Gamma)$ where Γ represents premise of the statement (i.e., if body-type is SUV or coupe) and $\mathcal{O}$ denotes the conclusion (i.e., Ford is preferred to Toyota). We define *primitive* preference for an atomic module as follows.

Definition 4. *Let M be an atomic module and $vocab(M) = \tau$. A τ_o-preference (or simply called preference) $\mathcal{P} = (\mathcal{O}, \Gamma)$ in M is a pair where $\mathcal{O} = (\mathcal{S}, \prec)$ is a strict partial order over $\mathcal{S}$ that is the set of all τ_o-partial structures in M where $\tau_o \subseteq \tau$. As well, $\Gamma = \{\gamma_1, \gamma_2, ..., \gamma_m\}$ is a set of τ_{p_i}-partial structures, $1 \leq i \leq m$, in M that $\tau_{p_i} \subseteq \tau$.*

In the above definition, Γ is a set of partial structures ($\Gamma = \{$SUV, coupe$\}$) and $\mathcal{O} = \{$Ford $\succ$ Toyota$\}$. Binary relation $\succ_\mathcal{P}$ is defined as follows:

Definition 5. *Let M be an atomic module, and $\mathcal{B}, \mathcal{B}'$ be two structures in M. Given a τ_o-preference $\mathcal{P} = (\mathcal{O}, \Gamma)$ in M, let Δ be a set of all structures in M that extend at least one member of Γ. We say structure $\mathcal{B}$ is preferred to $\mathcal{B}'$ with respect to $\mathcal{P}$ (denoted by $\mathcal{B} \succ_\mathcal{P} \mathcal{B}'$) if 1) $\mathcal{B}, \mathcal{B}' \in \Delta$, 2) there are partial structures $\mathcal{B}_i$ and $\mathcal{B}_j$ over $vocab(M)$ that can be extended* to structures in M such that $\mathcal{B}_i \succ \mathcal{B}_j$, and $\mathcal{B}$ is an extension of $\mathcal{B}_i$, where $\mathcal{B}'$ extends $\mathcal{B}_j$, and 3) there are no partial structures $\mathcal{B}_k$ and $\mathcal{B}_m$ such that $\mathcal{B}$ and $\mathcal{B}'$ extend them respectively and $\mathcal{B}_m \succ \mathcal{B}_k$.

This definition states that $\mathcal{B}$ is preferred to $\mathcal{B}'$ with respect to $\mathcal{P}$ if $\mathcal{B}_{\tau_o} \succ \mathcal{B}'_{\tau_o}$ with respect to $\mathcal{O}$ and both $\mathcal{B}$ and $\mathcal{B}'$ agree on at least a member of Γ ($\mathcal{B}|_{vocab(\gamma_r)} = \mathcal{B}'|_{vocab(\gamma_r)} = \gamma_r$ where $\gamma_r \in \Gamma$). It makes no difference how the rest of the vocabulary ($vocab(M) - \tau_o$) is interpreted because it is irrelevant to $\mathcal{P}$.

Example 3. *In Example 1, consider that safety of delivering items is an important preference for the company. So, it is preferred to avoid packing heavy and light items together to reduce the risk of damage to the light items. Each item i with its attributes (price, weight, and capacity) are associated with structure $Item_i$. Let $\mathcal{P}_{safe} = (\mathcal{O}_{safe}, \Gamma_{safe})$ be the safety preference where $\mathcal{O}_{safe} = (\mathcal{S}, \prec)$ is a partial order over $\mathcal{S}$ that is the set of all structures ($Item_i s$). Relation "$\prec$" is defined as $\{pack^-(i) \prec pack(i) | w(i) \leq W\}$; it means that for each item i that is lighter than a constant weight W, it is preferred to not put i in the pack. According to Notation 1, $pack^-(i)$ is representation of a partial structure that interprets ground atom $pack(i)$ as false. The premise of the conditional statement is formalized by $\Gamma_{safe} = \{\mathcal{C}_1, \mathcal{C}_2, ..., \mathcal{C}_m\}$ where $\mathcal{C}_i = \{item(i), w(i)\}$ such that $w(i) \geq W'$. This states when there is an item with weight not less than W', it is preferred to not include items lighter than W in the pack.*

POS preferences can be represented by $\mathcal{P}$ in our formalism. The following example shows how a POS is translated to a preference $\mathcal{P}$.

Example 4. *Imagine that we want to buy a car from a dealership. Let's assume POS set includes BMW and the POS preference indicates BMW cars are preferred to others. In the context of modular systems, we consider a module M that specifies a set of available cars in the dealership. Each structure of the module characterizes properties of a set of cars such as model, color, and body-type. Suppose Model is an unary relation (predicate) that indicates a car model, e.g. $Model(Benz)$. Assume $\mathcal{P} = (\mathcal{O}, \Gamma)$ where $\mathcal{O} = (\mathcal{S}, \prec)$ is a strict partial order over car properties such that $\prec$ is defined as: $Model(BMW) \succ Model(x)$, for all $x \neq BMW$, or simply $Model(BMW) \succ Model^-(BMW)$. According to Notation 1, this statements means that all structures that interpret Model as Model(BMW) are preferred. If car A is specified as A={Model(BMW), Color(white), Body-type(sedan)} and for car B we have B={Model(Benz), Color(Black), Body-type(SUV)}, then car A is preferred to car B through binary relation $\prec_\mathcal{P}$ that is equivalent with the the semantics of the POS preference. Considering a similar argument, NEG constructor can also be expressed in our language.*

The notions of two structures are *equally preferred* or *incomparable* are defined in the following.

Definition 6. *For two structures $\mathcal{B}, \mathcal{B}' \in M$, if a) neither $\mathcal{B} \succ_\mathcal{P} \mathcal{B}'$ nor $\mathcal{B}' \succ_\mathcal{P} \mathcal{B}$, b) for any $\mathcal{B}'' \in M$, if $\mathcal{B}'' \succ_\mathcal{P} \mathcal{B}$ then $\mathcal{B}'' \succ_\mathcal{P} \mathcal{B}'$, and c) if $\mathcal{B} \succ_\mathcal{P} \mathcal{B}''$ then $\mathcal{B}' \succ_\mathcal{P} \mathcal{B}''$, they are called equally preferred with respect to $\mathcal{P}$ and are represented by $\mathcal{B} \approx_\mathcal{P} \mathcal{B}'$. If one of the conditions (b) or (c) does not hold, then, $\mathcal{B}$ and $\mathcal{B}'$ are incomparable and are represented by $\mathcal{B} \sim_\mathcal{P} \mathcal{B}'$. Also, $\mathcal{B} \succeq_\mathcal{P} \mathcal{B}'$ means that $\mathcal{B} \succ_\mathcal{P} \mathcal{B}'$ or $\mathcal{B} \approx_\mathcal{P} \mathcal{B}'$.*

The following results are concluded from Definition 4, 5 and 6.

Proposition 1. *Given a preference $\mathcal{P} = (\mathcal{O}, \Gamma)$ in a module M, $\prec_\mathcal{P}$ is a strict partial order, $\approx_\mathcal{P}$ is an equivalence relation over structures of M, and $\succeq_\mathcal{P}$ is a transitive and reflexive binary relation over structures of M.*

Proof: As it has been specified in Definition 3, a binary relation is a strict partial order if it is anti-reflexive, asymmetrical, and transitive. Let $\mathcal{B}$ be a structure in M and $\mathcal{P} = (\mathcal{O}, \Gamma)$ be a τ_o-preference where $\tau_o \subseteq vocab(M)$. Assume that $\mathcal{B} \succ_\mathcal{P} \mathcal{B}$. According to Definition 5, there are partial structures $\mathcal{B}_i$ and $\mathcal{B}_j$ over $vocab(M)$ that can be extended to $\mathcal{B}$ such that $\mathcal{B}_i \succ \mathcal{B}_j$. Since both $\mathcal{B}_i$ and $\mathcal{B}_j$ are τ_o-partial structures, and they can be extended to the same total structure $\mathcal{B}$, we can conclude that $\mathcal{B}_i = \mathcal{B}_j$. An immediate consequence is $\mathcal{B}_i \succ \mathcal{B}_i$. This is impossible because $\mathcal{O}$ is a strict partial order. Therefore, it can be concluded that $\succ_\mathcal{P}$ is anti-reflexive. To prove asymmetry property, assume two structures $\mathcal{B}, \mathcal{B}' \in M$ and $\mathcal{B} \succ_\mathcal{P} \mathcal{B}'$. From Definition 5, there exist partial structures $\mathcal{B}_i$ and $\mathcal{B}_j$, that can be extended to $\mathcal{B}$ and $\mathcal{B}'$ respectively that are ordered as $\mathcal{B}_i \succ \mathcal{B}_j$. If $\succ_\mathcal{P}$ is symmetrical then we have $\mathcal{B}' \succ_\mathcal{P} \mathcal{B}$ that implies there exists partial structure $\mathcal{B}_m$, extendible to $\mathcal{B}$, and $\mathcal{B}_n$, extendible to $\mathcal{B}'$, over $vocab(M)$ such that $\mathcal{B}_n \succ \mathcal{B}_m$. With respect to Definition 5, we know that $\mathcal{B}_i, \mathcal{B}_j, \mathcal{B}_m, \mathcal{B}_n$ are τ_o-partial structures, where $\tau_o \subseteq vocab(M)$. Since $\mathcal{B}_i$ and $\mathcal{B}_n$ are extendible to the same total structure $\mathcal{B}$, immediate result is $\mathcal{B}_i = \mathcal{B}_n$. Having the same argument, we conclude $\mathcal{B}_j = \mathcal{B}_m$. Consequently, we have $\mathcal{B}_i \succ \mathcal{B}_j$ and at the same time $\mathcal{B}_j \succ \mathcal{B}_i$. This is a contradiction, so binary relation $\prec_\mathcal{P}$ is not symmetrical. It is straightforward to prove that the relation $\prec_\mathcal{P}$ is transitive. Let $\mathcal{B}, \mathcal{B}'$, and $\mathcal{B}''$ be total structures in M. Assume that $\mathcal{B} \succ_\mathcal{P} \mathcal{B}'$ and $\mathcal{B}' \succ_\mathcal{P} \mathcal{B}''$. Definition 5 states that there exist partial structures $\mathcal{B}_i, \mathcal{B}_j, \mathcal{B}_m, \mathcal{B}_n$ over $vocab(M)$ that $\mathcal{B}$ is an extension of $\mathcal{B}_i$, $\mathcal{B}_j$ and $\mathcal{B}_m$ are extendible to $\mathcal{B}'$, and $\mathcal{B}_n$ can be extended to $\mathcal{B}''$. Also we have $\mathcal{B}_i \succ \mathcal{B}_j$ and $\mathcal{B}_m \succ \mathcal{B}_n$. Considering the fact that partial structures $\mathcal{B}_j$ and $\mathcal{B}_m$ can be extended to the same total structure, it can be concluded that they are same. Consequently, we have $\mathcal{B}_i \succ \mathcal{B}_j$ and $\mathcal{B}_j \succ \mathcal{B}_n$. Since $\mathcal{O}$ is a strict partial order, we have $\mathcal{B}_i \succ \mathcal{B}_n$ that results in $\mathcal{B} \succ_\mathcal{P} \mathcal{B}''$. Consequently, $\prec_\mathcal{P}$ is a strict partial order.

To show that $\approx_\mathcal{P}$ is an equivalence relation, assume a structure $\mathcal{B}$. According to Definition 6, the relation $\approx_\mathcal{P}$ holds because $\succ_\mathcal{P}$ is not true due to the fact that $\prec_\mathcal{P}$ is a strict partial order. So, $\approx_\mathcal{P}$ is a reflexive relation. Similarly, according to Definition 6 condition a, $\approx_\mathcal{P}$ is symmetric. Also, conditions b and c guarantee that $\approx_\mathcal{P}$ is transitive. Therefore, $\approx_\mathcal{P}$ is an equivalence relation. Since $\succeq_\mathcal{P}$ means that $\succ'_\mathcal{P}$ or $\approx_\mathcal{P}$, it is clear that $\succeq_\mathcal{P}$ is transitive because both $\approx_\mathcal{P}$ and $\mathcal{B} \succ_\mathcal{P} \mathcal{B}'$ are transitive. Also, $\succeq_\mathcal{P}$ is reflexive since $\approx_\mathcal{P}$ is reflexive. However, $\succeq_\mathcal{P}$ is not guaranteed to be symmetric because $\approx_\mathcal{P}$ is not symmetric. So, $\succeq_\mathcal{P}$ is a reflexive and transitive (pre-order) relation. ∎

Meta-Preferences

In practice, each module may have more than one preference. Some of them may be preferred to others. For example, consider the following scenario: a transportation planning system decides one of three ways of transportation: walk, taxi, and bus. Let preference over time of travel be $\mathcal{P}_t$, and let preference over cost of travel be $\mathcal{P}_c$. Assume that a decision maker considers time more important than cost. It means that $\mathcal{P}_t$ is preferred to $\mathcal{P}_c$. The question then arises how a preferred transportation plan is identified in this case. The notion of meta-preference addresses this question in our framework.

Definition 7. *Given a module M and a set of preferences $\Pi = \{\mathcal{P}_1, \mathcal{P}_2, ..., \mathcal{P}_n\}$, let $\Omega_i := \{\mathcal{P}_j \in \Pi \mid (\mathcal{P}_i \succ \mathcal{P}_j) \vee (\mathcal{P}_j \succ \mathcal{P}_i)\}$ be a subset of Π such that its elements have order relation with $\mathcal{P}_i$. Assume $\mathcal{O}_{MP} = (\Pi, \prec)$ is a strict partial order over elements of Π. Binary relation $\succ_{\mathcal{MP}^*}$ over structures of M is defined as:*

$\mathcal{B} \succ_{\mathcal{MP}^*} \mathcal{B}'$ if there is a preference $\mathcal{P}_i \in \Pi$ such that $\mathcal{B} \succ_{\mathcal{P}_i} \mathcal{B}'$ and
- *there does not exist $\mathcal{P}_j \in \Omega_i$ that $\mathcal{P}_j \succ \mathcal{P}_i$ with respect to $\mathcal{O}_{MP}$ and $\mathcal{B}' \succ_{\mathcal{P}_j} \mathcal{B}$, and*
- *there is not a preference $\mathcal{P}_k \in \Pi \setminus \Omega_i$ that $\mathcal{B}' \succ_{\mathcal{P}_k} \mathcal{B}$.*

Meta-preference $\mathcal{MP}$ is characterized as $\mathcal{MP} := \mathcal{O}_{MP}$. We say structure $\mathcal{B}$ is preferred to $\mathcal{B}'$ with respect to binary relation $\succ_{\mathcal{MP}} \subseteq M \times M$ (notation $\mathcal{B} \succ_{\mathcal{MP}} \mathcal{B}'$) whenever if $\exists \mathcal{B}'' \in M; \mathcal{B}' \succ_{\mathcal{MP}^*} \mathcal{B}''$, then $\mathcal{B}'' \not\succ_{\mathcal{MP}^*} \mathcal{B}$.

This definition states that structure $\mathcal{B}$ is preferred to $\mathcal{B}'$ with respect to $\mathcal{MP}$ if we can find a preference such as $\mathcal{P}_i$ that $\mathcal{B} \succ_{\mathcal{P}_i} \mathcal{B}'$ and there is no preference that makes $\mathcal{B}'$ preferred to $\mathcal{B}$. If there is a preference $\mathcal{P}_j$ such that $\mathcal{B}' \succ_{\mathcal{P}_j} \mathcal{B}$ then $\mathcal{B}$ is not preferred to $\mathcal{B}'$ with respect to the meta-preference unless $\mathcal{P}_i$ is preferred to $\mathcal{P}_j$. Also, the definition prevents conflicts may happen between a mix of preferences, though it does not guarantee transitivity of $\succ_{\mathcal{MP}}$. If $\mathcal{B}$ is preferred to $\mathcal{B}'$ with respect to $\mathcal{MP}$, and if $\mathcal{B}'$ is preferred to $\mathcal{B}''$, then $\mathcal{B}''$ cannot be preferred to $\mathcal{B}$ with respect to $\mathcal{MP}$.

Example 5. *In Example 1, assume that the company has more than one preference. If an expensive item is selected for delivery, it is not secure to have another precious item in the pack that is specified by $\mathcal{P}_{security}$. Assume we have a meta-preference $\mathcal{MP}$ such that $\Pi_K = \{\mathcal{P}_{safe}, \mathcal{P}_{security}\}$ and $\mathcal{MP} = \{\mathcal{P}_{safe} \prec \mathcal{P}_{security}\}$. To have a preferred packing for the Knapsack module, when there is a heavy and expensive item in the pack, it is preferred to not include another heavy item, but it is fine to have two expensive items in the pack.*

Preference-based Modular Systems

Up to now, we defined *primitive* preference $\mathcal{P}$ in atomic modules. In what follows, we study how *compound* preference in a modular system is constructed when preferences of its components are given. First, we define *product* of preferences.

Definition 8. *Assume* $M = M_1 \times M_2$ *is a modular system such that* $vocab(M_1) = \tau_1$ *and* $vocab(M_2) = \tau_2$. *Let* $\mathcal{P}_1$ *and* $\mathcal{P}_2$ *be (primitive or compound) preferences in* M_1 *and* M_2 *respectively. We say that* $\mathcal{P}$ *is product of* $\mathcal{P}_1$ *and* $\mathcal{P}_2$ *(notation* $\mathcal{P} = \mathcal{P}_1 \times \mathcal{P}_2$*) and extend* $\succ_\mathcal{P}$ *to products by what follows. Let's consider* $\mathcal{B}, \mathcal{B}' \in M$. *Structure* $\mathcal{B}$ *is preferred to structure* $\mathcal{B}'$ *(denoted by* $\mathcal{B} \succ_{\mathcal{P}_1 \times \mathcal{P}_2} \mathcal{B}'$*) if and only if* $\mathcal{B}|_{\tau_i} \succ_{\mathcal{P}_i} \mathcal{B}'|_{\tau_i}$ *for* $i \in \{1, 2\}$.

Informally, $\mathcal{B}$ is preferred to $\mathcal{B}'$ with respect to $\mathcal{P} = \mathcal{P}_1 \times \mathcal{P}_2$, if $\mathcal{B}$ is preferred to $\mathcal{B}'$ with respect to $\mathcal{P}_1$ when they are restricted to the vocabulary of M_1 and with respect to $\mathcal{P}_2$ when they are restricted to the vocabulary of M_2.

Example 6. *In Example 1, for module* M_{tsp}, *suppose that if cities* c_1, c_2, c_3, c_4 *are in the set of destinations, there is a path from* c_1 *to* c_4 *through* c_2 *that is preferred to the path from* c_1 *to* c_4 *through* c_3. *This can be formalized by preference* $\mathcal{P}_{tsp} = (\mathcal{O}_{tsp}, \Gamma_{tsp})$ *where* $\mathcal{O}_{tsp} = (\mathcal{S}_{tsp}, \prec)$ *is a partial order over* $\mathcal{S}_{tsp}$ *that is the set of all possible routes. For a positive integer k and truck t,*

$\{Route(k, c_1, t), Route(k+1, c_2, t), Route(k+2, c_4, t) \succ$
$Route(k, c_1, t), Route(k+1, c_3, t), Route(k+2, c_4, t)\}$ *and*
$\Gamma_{tsp} = \{Dest(1, c_1), Dest(2, c_2), Dest(3, c_3), Dest(4, c_4)\}$.

A preferred plan of packing and delivery with respect to $\mathcal{P}_{safe} \times \mathcal{P}_{tsp}$ *is the one where heavy and light items are not in the same pack and if the truck is supposed to visit cities* c_1, c_2, c_3, c_4, *then taking road* (c_1, c_2) *is preferred to* (c_1, c_3).

Given $\mathcal{MP}_1$ on $\Pi_1 = \{\mathcal{P}_{11}, ..., \mathcal{P}_{1r}\}$ in M_1 and $\mathcal{MP}_2$ on $\Pi_2 = \{\mathcal{P}_{21}, ..., \mathcal{P}_{2s}\}$ in M_2, we introduce binary relation $\mathcal{MP} = \mathcal{MP}_1 \times \mathcal{MP}_2$ on $\Pi = \{\mathcal{P}_1, ..., \mathcal{P}_{r \times s}\}$ where $\mathcal{P}_i = \mathcal{P}_{1i} \times \mathcal{P}_{2i}$ as follows:

Definition 9. *Binary relation* $\succ_{\mathcal{MP}}$ *is defined as* $\{(\mathcal{A}, \mathcal{B}) | \mathcal{A}, \mathcal{B} \in M_1 \times M_2 \text{ and } \mathcal{A}|_{\tau_1} \succ_{\mathcal{MP}_1} \mathcal{B}|_{\tau_1} \text{ and } \mathcal{A}|_{\tau_2} \succ_{\mathcal{MP}_2} \mathcal{B}|_{\tau_2}\}$.

A *compound* preference can be constructed by union of two preferences as follows:

Definition 10. *Let* $M = M_1 + M_2$ *be a modular system. Suppose* $vocab(M_1) = \tau_1$ *and* $vocab(M_2) = \tau_2$. *Assume* $\mathcal{P}_1$ *and* $\mathcal{P}_2$ *are preferences in* M_1 *and* M_2 *respectively. For* $\mathcal{B}, \mathcal{B}' \in M$, *if* $\mathcal{B}|_{\tau_1} \succ_{\mathcal{P}_1} \mathcal{B}'|_{\tau_1}$ *or* $\mathcal{B}|_{\tau_2} \succ_{\mathcal{P}_2} \mathcal{B}'|_{\tau_2}$ *then* $\mathcal{B}$ *is preferred to* $\mathcal{B}'$ *with respect to* $\mathcal{P}_1 \cup \mathcal{P}_2$ *and is denoted by* $\mathcal{B} \succ_{\mathcal{P}_1 + \mathcal{P}_2} \mathcal{B}'$.

For meta-preferences, similar to Definition 9, given $\mathcal{MP} = \mathcal{MP}_1 \cup \mathcal{MP}_2$, binary relation $\succ_{\mathcal{MP}}$ is $\{(\mathcal{A},\mathcal{B})|\mathcal{A},\mathcal{B} \in M_1 \times M_2 \text{ and } \mathcal{A}|_{\tau_1} \succ_{\mathcal{MP}_1} \mathcal{B}|_{\tau_1} \text{ or } \mathcal{A}|_{\tau_2} \succ_{\mathcal{MP}_2} \mathcal{B}|_{\tau_2}\}$.

Selection operator can be applied on preference $\mathcal{P}'$ in module M' that is denoted by $\mathcal{P} = \sigma_{Q \equiv R}(\mathcal{P}')$. Let us comment briefly on the selection operator. Assume M is a τ-modular system and $R,Q \in \tau$. Selection does not change the vocabulary of M. Hence, definition of preference remains unchanged. Given a preference $\mathcal{P}$ in M, when $\mathcal{B} \succ_{\mathcal{P}} \mathcal{B}'$ holds in M, if $\mathcal{B}$ and $\mathcal{B}'$ are also structures of $\sigma_{R \equiv Q} M$, we can conclude that $\mathcal{B}$ is preferred to $\mathcal{B}'$ in $\sigma_{R \equiv Q} M$.

Definition 11. *Let's assume $M' = \sigma_{R \equiv Q} M$ and $\mathcal{P} = (\mathcal{O}, \Gamma)$ is a preference in M. We introduce compound preference $\mathcal{P}' = \sigma_{Q \equiv R} \mathcal{P}$. For $\mathcal{B},\mathcal{B}' \in M$, whenever $\mathcal{B} \succ_{\mathcal{P}} \mathcal{B}'$, if $\mathcal{B},\mathcal{B}' \in M'$ then $\mathcal{B} \succ_{\mathcal{P}'} \mathcal{B}'$.*

This definition says that if two structures $\mathcal{B}$ and $\mathcal{B}'$ are in M, and $\mathcal{B}$ is preferred to $\mathcal{B}'$ with respect to $\mathcal{P}$ then $\mathcal{B}$ remains preferred to $\mathcal{B}'$ in module M' that is module M with selection operator. For module M with meta-preference $\mathcal{MP}$ over preferences Π, binary relation $\succ_{\sigma_{R \equiv Q} \mathcal{MP}} = \{(\mathcal{A},\mathcal{B})|(\mathcal{A},\mathcal{B}) \in M, M' \text{ and } \mathcal{A} \succ_{\mathcal{MP}} \mathcal{B}\}$ is defined.

Projection is another operator that can be used to build *compound* preferences. Projection operator hides some vocabulary symbols so it restricts number of models. Intuitively, given two structures $\mathcal{A}'$ and $\mathcal{B}'$ in $M' = \pi_\delta(M)$, if all structures in M which their projection to δ are equal to $\mathcal{A}'$, are preferred to all structures in M that their projection to δ are equal to $\mathcal{B}'$, we can conclude $\mathcal{A}'$ is preferred to $\mathcal{B}'$. The following definition formulates projected preference.

Definition 12. *Assume $M' = \pi_\delta(M)$ where $vocab(M) = \tau$ and $vocab(M') = \delta$. Let $\mathcal{P}$ be a preference in M. We define $\mathcal{P}' = \pi_\delta(\mathcal{P})$ and binary relation $\succ_{\mathcal{P}'} = \{(\mathcal{A}',\mathcal{B}')|\mathcal{A}',\mathcal{B}' \in M' \wedge \forall \mathcal{A} \in M; \mathcal{A}|_\delta = \mathcal{A}' \Rightarrow (\forall \mathcal{B}; \mathcal{B}|_\delta = \mathcal{B}' \Rightarrow \mathcal{A} \succ_{\mathcal{P}} \mathcal{B})\}$.*

Analogously, we introduce the concept projected meta-preference $\mathcal{MP}'$ in M'. Binary relation $\succ_{\pi_\delta(\mathcal{MP})}$ is defined as $\{(\mathcal{A}',\mathcal{B}')|\mathcal{A}',\mathcal{B}' \in M' \wedge \forall \mathcal{A} \in M; \mathcal{A}|_\delta = \mathcal{A}' \Rightarrow (\forall \mathcal{B}; \mathcal{B}|_\delta = \mathcal{B}' \Rightarrow \mathcal{A} \succ_{\mathcal{MP}} \mathcal{B})\}$.

The last operator for constructing *compound* preferences is complement. Similar to selection operator, vocabulary of module $M' = -M$ is not changed ($vocab(M) = vocab(M)'$). So, preferences in M' are the same as preferences in M.

Definition 13. *Let's assume $M' = -M$ be a modular system and $\mathcal{P}$ be a preference in M. Binary relation $\succ_{-\mathcal{P}}$ is defined as $\{(\mathcal{A},\mathcal{B})|\mathcal{A},\mathcal{B} \notin M \text{ and } \mathcal{A} \succ_{\mathcal{P}} \mathcal{B}\}$. Similarly for meta-preference $\mathcal{MP}$ in M, we define $\succ_{-\mathcal{MP}}$ as $\{(\mathcal{A},\mathcal{B})|\mathcal{A},\mathcal{B} \notin M \text{ and } \mathcal{A} \succ_{\mathcal{MP}} \mathcal{B}\}$.*

We introduce the notion of preference-based modular system $\mathcal{P}$-$\mathcal{MS}$ that is a modular system with a partial order over its *compound* or *atomic* preferences.

Definition 14. *A modular system MS with a set of compound or atomic preferences Π is a preferred modular system, notation $\mathcal{P}$-$\mathcal{MS}$, if it is specified by a pair $(\prec_{\mathcal{MP}}, MS)$ where $\mathcal{MP}$ is a meta-preference in MS.*

The following result is a major property of our framework that is *closed* in the sense that the operations for combining modules preserve the notion of a preference-based modular system. This property is proven by structural induction.

Theorem 1. *Assume for some n, a modular system MS is obtained from $M_1, M_2, ..., M_n$, where $M_i s$, $1 \leq i \leq n$ are modular systems, by using operations in modular systems including product, union, complement, selection, and projection. For all $1 \leq i \leq n$, if M_i is $\mathcal{P}$-$\mathcal{MS}$ then M is also $\mathcal{P}$-$\mathcal{MS}$.*

Proof: A preference-based modular system ($\mathcal{P}$-$\mathcal{MS}$) is a pair $(\succ_{\mathcal{MP}}, MS)$ where $\succ_{\mathcal{MP}}$ is meta-preference over preferences on vocabulary of MS. Assume that $M = M_1 \circledast M_2 \circledast ... \circledast M_n$ where M_i $i \in \{1..., n\}$ is a preference-based modular system and $\circledast$ is an operator of modular systems. We prove by induction that M is a $\mathcal{P}$-$\mathcal{MS}$. For $n = 1$, it is trivial that MS is a $\mathcal{P}$-$\mathcal{MS}$. Let's assume $M' = M_1 \circledast M_2$. According to Definition 14, M' is a modular system. Assume $\circledast$ is product operator (for other operators we use the same argument). So, $vocab(M') = vocab(M_1) \cup vocab(M_2)$, and $\mathcal{MP}' = \mathcal{MP}_1 \times \mathcal{MP}_2$ is meta-preference over $vocab(M')$. Therefore, M' is a modular system with meta-preference over its vocabulary. Thus, M' is a $\mathcal{P}$-$\mathcal{MS}$. We can rewrite $M = M' \circledast M_3 ... \circledast M_n$. Having the same argument, if $M' = M_1 \circledast M_3 ... \circledast M_k$, $k < n$, a $\mathcal{P}$-$\mathcal{MS}$, then $M' \circledast M_{k+1}$ is $\mathcal{P}$-$\mathcal{MS}$. Consequently, MS is a preference-based modular system.

∎

4 Relationship of $\mathcal{P}$-$\mathcal{MS}$ to other Preference Formalisms

We now describe two preference formalisms and show how they can be related to our formalism.

4.1 CP-Nets

Ceteris paribus (cp) network is a graphical representation of conditional preferences with reasoning capability [5]. The idea underlying cp-nets is to compare different assignments to a set of variables as some of these variables are conditionally dependent on each other. Each node represents an attribute (variable) connected to its parents through directed edges. A preference order over domain values of a variable is dependent on value of its parent variables. The dependency is shown by a *Conditional Preference Table* (CPT) represented as an annotation for each node. There exists an induced graph derived from each cp-net that shows ordering relation between a subset of outcomes. Each node in the induced graph represents an outcome and each directed edge exhibits ordering relation between nodes. An outcome o_1 is preferred to o_2 if in the induced graph, there is a path from o_1 to o_2. An induced graph comprises all information about preferences over outcomes that can be derived from a cp-net.

From the syntactic point of view, $\mathcal{P}\text{-}\mathcal{MS}$ is able to capture the notion of attributes in cp-nets. Each attribute can be viewed as an interpreted predicate symbol in the context of $\mathcal{P}\text{-}\mathcal{MS}$. Therefore, an outcome in a cp-net can be represented by a structure that interprets vocabulary symbols. The relation between cp-nets and $\mathcal{P}\text{-}\mathcal{MS}$ in this way implies that the space of all outcomes in a cp-net can be modelled by a set of structures interpreting vocabulary symbols in $\mathcal{P}\text{-}\mathcal{MS}$. A preference statement visualized by a cp-net over a set of variables $V = \{V_1, ..., V_n\}$ is an ordering over domain values of a variable that may or may not be dependent on some other variables, and a preference in $\mathcal{P}\text{-}\mathcal{MS}$ is defined as $\mathcal{P} = (\mathcal{O}, \Gamma)$ where $\mathcal{O}$ is a partial order given a set of partial structures Γ. In a sense, a partial structure in $\mathcal{P}\text{-}\mathcal{MS}$ is a combination of some interpreted vocabulary symbols. Thus, a partial structure can stipulate a value assigned to an attribute. Orderings over partial structures in our formalism are in fact orderings over attribute values in cp-nets when partial structures in $\mathcal{O}$ are assumed to interpret only one vocabulary symbol. Transforming the condition part of the preference statement in a cp-net is straightforward. Order relation holds for partial structures which extend Γ. Therefore, parents of each cp-net attribute can be represented by Γ.

In order to establish the correspondence between the semantic of cp-nets and $\mathcal{P}\text{-}\mathcal{MS}$, first we explain the concept of *flip-over* in cp-nets. In an induced graph derived from a cp-net, each outcome node has one attribute value preferred to its child's while other attributes are assumed to be fixed. Therefore, by moving from a node to its children one attribute value is changed that is called a *flip-over*. A path in an induced graph is a chain of *flip-overs* between two outcomes. Hence, an outcome is preferred to another when single or multiple *flip-over(s)* exist between them. Now, we show how a *flip-over* can be represented in $\mathcal{P}\text{-}\mathcal{MS}$. Consider two structures $\mathcal{B}$ and $\mathcal{B}'$; if $\mathcal{B} \succeq_{\mathcal{MP}} \mathcal{B}'$ ($\succeq_{\mathcal{MP}}$ means that $\succ_{\mathcal{MP}}$ or $\approx_{\mathcal{MP}}$ that is an equivalence relation), we have enough information to know that $\mathcal{B}$ is preferred to $\mathcal{B}'$ at least at one vocabulary symbol interpretation or they are equally preferred. The concept of a single *flip-over* can be specified by $\succeq_{\mathcal{MP}}$ when $\mathcal{O}_{MP} = \emptyset$ (there is no meta-preference in cp-nets). In this case, $\succeq_{\mathcal{MP}}$ has the transitivity property and a chain of *flip-overs* can be modelled by $\mathcal{P}\text{-}\mathcal{MS}$ as well. If $\mathcal{O}_{MP}$ is not empty, $\mathcal{MP}$ represents the notion of *relative importance* (meta-preference) in TCP-net [7] that is an extension of cp-nets to model meta-preferences. This reasoning leads us to the following theorem, relating cp-nets and the $\mathcal{P}\text{-}\mathcal{MS}$ formalism.

Theorem 2. *Let $\mathcal{N}$ be a cp-net and $\mathcal{MP}$ be the representation of $\mathcal{G}$ in the context of $\mathcal{P}\text{-}\mathcal{MS}$. If an outcome o_i is preferred to outcome o_j in the induced graph $\mathcal{G}$ of $\mathcal{N}$, then, for o_i and o_j that are transformed into the $\mathcal{P}\text{-}\mathcal{MS}$, we have $o_i \succeq_{\mathcal{MP}} o_j$.*

Proof: Theorem 2 Recall that a structure is a domain together with interpretation of a set of non-logical symbols. For a set of variables $X = \{X_1, X_2, ..., X_n\}$ and a domain $D = D(X_1) \times ... \times D(X_n)$, an outcome $o = \{x_1, ..., x_n\}$ in $\mathcal{G}$ is a structure $\mathcal{A}$ over vocabulary $\{X_1, X_2, ..., X_n\}$ such that $\mathcal{A} = (D, \mathcal{I})$ where $\mathcal{I}$ is an interpretation function such that $x_i \in D(X_i)$ and $x_i = \mathcal{I}(X_i)$, $i \in \{1, ..., n\}$. In graph $\mathcal{G}$, an outcome node o_i is preferred to its child o_j and they are all the same except in one variable called X_d. Consider in o_i, we have $X_d = x_{di}$ and for o_j, we have $X_d = x_{dj}$. Since o_i is preferred to o_j, the following holds: $x_{di} \succ x_{dj}$. Now, suppose $\mathcal{A}$ is a structure representing o_i and o_j is represented by $\mathcal{B}$. Consider two partial structures $\mathcal{A}_i$ and $\mathcal{B}_j$ such that $\mathcal{I}_{\mathcal{A}_i}(X_d) = x_{di}$ and $\mathcal{I}_{\mathcal{B}_j}(X_d) = x_{dj}$. Since $x_{di} \succ x_{dj}$, in $\mathcal{P}\text{-}\mathcal{MS}$ framework, there is a preference $\mathcal{P}$ that $\mathcal{A}_i \succ_\mathcal{P} \mathcal{B}_j$. According to Definition 5, we conclude that $\mathcal{A} \succ_\mathcal{P} \mathcal{B}$ such that $\mathcal{A}$ and $\mathcal{B}$ are representation of o_1 and o_2 respectively. If o_1 and o_2 are not adjacent nodes in $\mathcal{G}$ but there is a path from o_1 to o_2, by induction it is derived that $\mathcal{A} \succ_{\mathcal{MP}} \mathcal{B}$ where $\mathcal{O}_{MP} = \emptyset$ (there is no meta-preference in cp-nets). ∎

The following example shows how cp-nets can be expressed in $\mathcal{P}\text{-}\mathcal{MS}$.

Example 7. *Consider car dealership scenario in Example 4. Each car has three attributes including model, type, and color. Model can be Benz or BMW in white or black color. Each car can be in SUV or sedan type. Available cars in the dealership is listed in Table 1. It can be observed from Figure 2.a, that black car is always better than white one regardless of type or model. For a black color, Benz is preferred to BMW, and vice versa. Similarly, sedan BMW is preferred to SUV type. Conversely for Benz, it is preferred to buy SUV rather than sedan. The graph induced from this cp-net is illustrated in Figure 2.b. The induced graph shows whether a certain outcome is preferred to another one. For example outcome o_1 that is a black SUV Benz is preferred to o_6 that is white SUV BMW because there is a path form o_1 to o_6.*

To represent this cp-net in $\mathcal{P}\text{-}\mathcal{MS}$ formalism, we consider a primitive module (there is no combination of cp-nets) MS. Let's assume M is an unary relation (predicate) that indicates a car model, e.g. $M(Benz)$ or $M(BMW)$. Similarly, T specifies type, including $T(sedan)$ and $T(SUV)$. For color, predicate C can take value white and black ($C(white)$ and $C(black)$). Let $\mathcal{P}_1 = (\mathcal{O}_1, \Gamma_1)$ where $\mathcal{O} = (\mathcal{S}_1, \prec_1)$ is a strict partial order over color attribute such that $\prec_1$ is defined as: $C(black) \succ C(white)$, and $\Gamma_1 = \emptyset$. Infact, $\mathcal{P}_1$ represents the first row in Figure 2.a. In a similar way, $\mathcal{P}_2 = (\mathcal{O}_1, \Gamma_1)$ is defined as $\Gamma_2 = C(black)$ and $\mathcal{O}_2 = (\mathcal{S}_2, \prec_2)$ where $\mathcal{S}_2 = T(BMW) \succ T(Benz)$. We construct $\mathcal{P}_3$ and $\mathcal{P}_4$ likewise.

A SEMANTIC APPROACH TO COMBINING PREFERENCE FORMALISMS

Outcome	Model	Color	Type
o_1	Benz	Black	SUV
o_2	BMW	Black	SUV
o_3	Benz	White	SUV
o_4	Benz	Black	Sedan
o_5	Benz	White	Sedan
o_6	BMW	White	SUV
o_7	BMW	Black	Sedan
o_8	BMW	White	Sedan

Table 1: Outcomes in the car dealership example

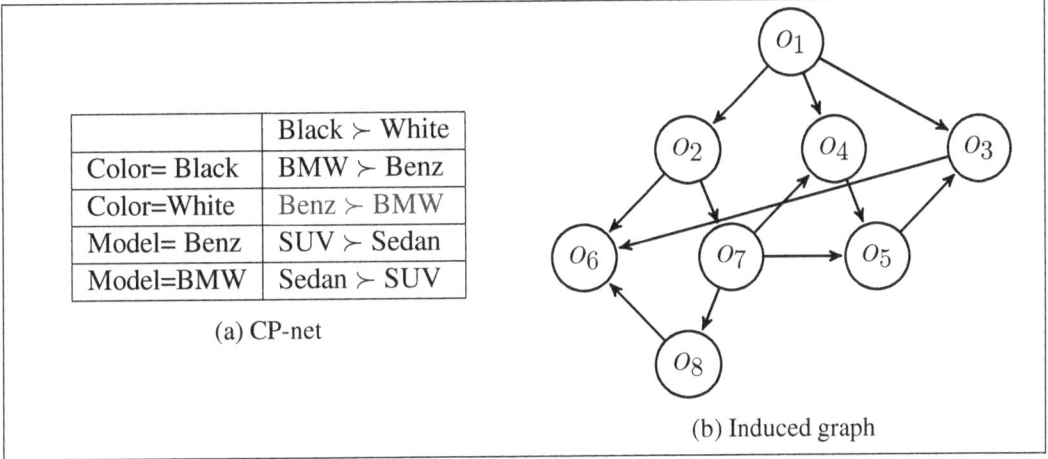

Figure 2: CP-net example: car dealership

For a set of preferences $\Pi = \{\mathcal{P}_1, \mathcal{P}_2, \mathcal{P}_3, \mathcal{P}_4\}$, meta-preference $\mathcal{O}_{MP} = (\Pi, \succ)$ is defined as $\succ = \emptyset$ because there is no preference over preferences. Assume structure $\mathcal{A}_i$ represents outcome o_i. According to Definition 7, it can be proven that $\mathcal{A}_i$ is preferred to $\mathcal{A}_j$ if and only if there is a path from o_i to o_j in Figure 2.b.

4.2 Preference-based Planning

In what follows, we show how $\mathcal{P}\text{-}\mathcal{MS}$ is able to assert preference statements expressed in $\mathcal{PP}$ [26] that is a preference language for planning problems. While we do not discuss the full details of $\mathcal{PP}$ here, we recall the main definitions found in [26]. Given a set of fluent symbols $\mathcal{F}$ and a set of actions $\mathcal{A}$, a state is defined as a subset of $\mathcal{F}$. A planning problem is a triple $\langle D, I, G \rangle$ where D indicates pre-conditions and effects of actions, I is the initial state, and G stands for the goal state. A solution to a planing problem, that is called a plan, is a chain of actions and states $I, a_1, ...a_n, G$ that starts from I and ends to G. A *basic desire* ϕ is identified as one of the following: 1) a certain action occurs in the plan denoted by $\phi \equiv occ(a)$, 2) a set of certain fluents are satisfied that is denoted by $\phi \equiv (f_i \wedge ... \wedge f_{i+n})$, 3) any combination of *basic desires* by using classical logic connectives (e.g. $\wedge$, $\vee$, and $\neg$) or temporal connective stemmed from temporal logic such as Next(ϕ_1), Until(ϕ_1, ϕ_2), Always(ϕ), and Eventually(ϕ).

[26] states that a planning problem $\langle D, I, G \rangle$ can be reduced to an Answer Set Programming (ASP) problem $\Pi(D, I, G)$ such that for a feasible plan p_M there is an answer set M in program Π. In the context of answer set programming, a formula ϕ is satisfied in M if it is a subset of vocabulary symbols that M makes true. For two plans p_1 and p_2, we say p_1 is preferred to p_2 with respect to a *basic desire* ϕ if ϕ is satisfied in p_1 but not in p_2. In the context of ASP, if M_1 and M_2 are two answer sets of p_1 and p_2 respectively, M_1 satisfies ϕ but M_2 does not.

To express *basic desires* in $\mathcal{P}$-$\mathcal{MS}$, it suffices to show that answer sets can be translated to structures in the context of modular systems. Consider a vocabulary $\{a_1, ..., a_n\}$ and an answer set $M = \{a_1, ..., a_k\}$ ($k \leq n$). As it can be observed from the notion of answer sets, M can be viewed as a structure that interprets each atom a_i, $i \leq k$, as true and for all a_j, $k < j \leq n$, as false. Having the same argument, a *basic desire* ϕ is a partial structure in modular systems such that a subset of atoms in M is true. As a result, a planning problem $\langle D, I, G \rangle$ with preferences can be translated into answer sets and then to modular systems. Assume that structure $\mathcal{B}$ represents plan p_1, structure $\mathcal{B}'$ is the translation of p_2, and formula ϕ is translated into partial structure $\mathcal{B}_\phi$. A plan p_1 is preferred to p_2 with respect to ϕ when $\mathcal{B}$ is preferred to $\mathcal{B}'$ with respect to $\mathcal{B}_\phi$. This completely coincides with our definition of preferences in modular systems. The following result follows from what we discussed.

Theorem 3. *Let p_1 and p_2 be two feasible plans of a planning problem $\langle D, I, G \rangle$ that can be translated to ASP program $\Pi(D, I, G)$. Let M_{p_1} and M_{p_2} be ASP translation of p_1 and p_2 respectively. Suppose that M_{p_1} is translated to structure $\mathcal{B}$ and M_{p_2} to structure $\mathcal{B}'$ in the context of $\mathcal{P}$-$\mathcal{MS}$. Given a basic desire ϕ, if p_1 is preferred to p_2 with respect to ϕ in language $\mathcal{PP}$, then $\mathcal{B} \succ_{\mathcal{MP}_\phi} \mathcal{B}'$ in $\mathcal{P}$-$\mathcal{MS}$ where $\mathcal{MP}_\phi$ is translation of ϕ into $\mathcal{P}$-$\mathcal{MS}$.*

Proof:Theorem 3 Without loss of generality, for a set of atoms $\tau = \{a_1, ..., a_n\}$, assume $M_{p_1} = \{a_m, ..., a_k\}$ where $1 \leq m \leq n$ and $m \leq k \leq n$, and $M_{p_2} = \{a_p, ..., a_q\}$ where $1 \leq p \leq n$ and $p \leq p \leq n$. Consider structure $\mathcal{A} = (\mathcal{I}, D)$ over τ such that $D = \{\text{True}, \text{False}\}$ and $\mathcal{I}(a_i) = $ True or False, $a_i \in M_{p_1}$. $\mathcal{A}$ is a presentation of M_{p_1}. Similarly, $\mathcal{B}$ represents M_{p_2}. Assume a basic desire ϕ that is a subset of τ and is defined as $\{a_{\phi_r}, ..., a_{\phi_s}\}$. Let $\mathcal{A}_i = \{a_{\phi_r}^+, ..., a_{\phi_s}^+\}$ be a τ_o-partial structure that satisfies ϕ. Also, let $\mathcal{B}_j$ be a τ_o-partial structure that falsifies ϕ. According to Notation 1, $\mathcal{B}_j$ has at least an atom in the form of $a_{\phi_t}^-$, $r \leq t \leq s$. Consider a preference $\mathcal{P} = (\Gamma, \mathcal{O})$ such that $\Gamma = \emptyset$ and $\mathcal{O} = \{\mathcal{A}_i \succ \mathcal{B}_j\}$. If

$\mathcal{A}$ extends $\mathcal{A}_i$ and $\mathcal{B}$ is an extension of $\mathcal{B}_j$, then $\mathcal{A} \succ_\mathcal{P} \mathcal{B}$. This means that structure $\mathcal{A}$ that satisfies ϕ is preferred to structure $\mathcal{B}$ that does not satisfy ϕ. Thus, a basic desire preference can be expressed in $\mathcal{P}\text{-}\mathcal{MS}$. ∎

Example 8. *Assume we want to travel from location l_1 to l_3 through l_2. Action Travel (notation T_1) from l_1 to l_2 can be done by taxi, bus, or train. Similarly, T_2 is travel from l_2 to l_3 by train or walk. Because of expensive cost, two consecutive travels cannot be completed by taking train. Also, walking is preferred to taking train or bus. Valid travel plans in $\mathcal{PP}$ are $p_1 = \{T_1(train), T_2(walk)\}$, $p_2 = \{T_1(bus), T_2(walk)\}$, $p_3 = \{T_1(bus), T_2(train)\}$, $p_4 = \{T_1(taxi), T_2(walk)\}$, and $p_5 = \{T_1(taxi), T_2(train)\}$. Preference is a basic desire that implies, for example, p_1 is preferred to p_3. Assume structures M_{p_2} and M_{p_3} are translation of p_2 and p_3 respectively in $\mathcal{P}\text{-}\mathcal{MS}$. For structure $M_{p_2} = (D, T_1^{M_{p_2}}, T_2^{M_{p_2}})$, $D = \{taxi, walk, bus, train\}$ is domain and $T_1^{M_{p_2}} = \{T_1(bus)\}$ and $T_2^{M_{p_2}} = \{T_2(walk)\}$ are interpretations of T_1 and T_2. Also, $M_{p_3} = (D, T_1^{M_{p_3}}, T_2^{M_{p_3}})$ where $D = \{taxi, walk, bus, train\}$, interprets T_1 as $T_1^{M_{p_3}} = \{T_1(bus)\}$ and T_2 as $T_2^{M_{p_3}} = \{T_2(train)\}$. Basic desire is expressed in $\mathcal{P}\text{-}\mathcal{MS}$ as a preference $\mathcal{P} = (\Gamma, \mathcal{O})$ where $\Gamma = \emptyset$ and $\mathcal{O} = \{T_1(walk)^+ \succ T_1(walk)^-, T_2(walk)^+ \succ T_2(walk)^-\}$. According to Definition 5, $M_{p_2} \succ_\mathcal{P} M_{p_3}$ that is exactly suggested by Theorem 3.*

4.3 Answer Set Optimization

Answer Set Optimization (ASO) is a framework to represent preferences in the context of Answer Set Programming [9]. An ASO program P is a pair (P_{gen}, P_{pref}) where P_{gen} is the generating program and P_{pref} is a set of rules in the form of
$r: C_1 > ... > C_k \leftarrow a_1, ..., a_n, not\ b_1, ..., not\ b_m$

In each rule, C_i is a combination of atoms integrated through classical logic connectives. Moreover, a_i and b_i are literals. Given a set of l rules $r_1, ..., r_l$, each model of P_{gen} is associated with a vector $V(M) = \{v_1(M), ..., v_l(M)\}$ where $v_i(M)$ is the rank of model M in r_i. The notion of rank denotes the minimum j of C_js in r_i that are satisfied by M. If M_1 and M_2 are two models of P_{gen}, M_1 is preferred to M_2 with respect to P_{pref} if $V(M_1) < V(M_2)$. In other words, suppose the right hand side (body) of the rule is satisfied by M_1 and M_2 while C_p and C_q are the most preferred C_is satisfied by M_1 and M_2 respectively. If C_p is preferred to C_q then M_1 is preferred to M_2. Also, when a model M does not satisfy body of a rule r_k, it is assumed that $v(M) = 1$.

We show the connection between ASO and $\mathcal{P}\text{-}\mathcal{MS}$. Assume $\{M_1, ..., M_n\}$ are models of an ASO program $P = (P_{gen}, P_{pref})$. Each model $M_i = \{l_1, l_2, ..., l_{m_i}\}$, generated by P_{gen}, is a set of literals. M_i can be viewed as a mathematical structure that its domain is formed from only two elements: true and false. So, P_{gen} is a modular system (a set of structures). Consider a preference rule $r: C_1 > ... > C_k \leftarrow a_1, ..., a_n, \text{not } b_1, ..., \text{not } b_m$. A partial structure $\mathcal{B}_i$ (or a set of partial structures in the case of logical combinations of literals) can represent each C_i. The body of the rule is a set of positive and negative literals that can be simply expressed by a partial structure as well. In the view of this observation, a preference in the form of $\mathcal{P} = (\Gamma, \mathcal{O})$ can assert a preference rule in ASO. The body of such rule is modelled by Γ and $\mathcal{O}$ that can represent the head of the rule. Each rule r_i can be represented by a preference $\mathcal{P}_i$. Therefore, a set of rules P_{pref} can be presented by meta-preference $\mathcal{MP}$.

To clarify the relation between ASO and $\mathcal{P}\text{-}\mathcal{MS}$, let us consider the food planner example discussed in [9]. Suppose a food menu consists of four sections: appetizer, main-food, drink, and dessert. The domain of each variable is defined as: $\mathcal{D}(\text{food}) = \{\text{fish, beef}\}$, $\mathcal{D}(\text{drink}) = \{\text{red-wine, white-wine, beer}\}$, $\mathcal{D}(\text{appetizer}) = \{\text{soup, salad}\}$, and $\mathcal{D}(\text{dessert}) = \{\text{ice-cream, pie}\}$. Let's assume the preference rules as below:

$r_1 = white\text{-}wine > red\text{-}wine > beer \leftarrow fish$
$r_2 = red\text{-}wine > beer > white\text{-}wine \leftarrow beef$
$r_3 = pie > ice\text{-}cream \leftarrow beer$

And consider a P_{gen} that generates the following models:

$M_1 = \{soup, beef, beer, ice\text{-}cream\}$
$M_2 = \{salad, beef, beer, ice\text{-}cream\}$
$M_3 = \{soup, fish, beer, ice\text{-}cream\}$
$M_4 = \{salad, fish, beer, ice\text{-}cream\}$

As noted above, the representation of r_1 in $\mathcal{P}\text{-}\mathcal{MS}$ is: $\mathcal{P}_1 = (\mathcal{O}_1, \Gamma_1)$ where $\mathcal{O}_1 = \{\text{drink(white-wine)} \succ \text{drink(red-wine)} \succ \text{drink(beer)}\}$ and $\Gamma_1 = \{\text{main-food(fish)}\}$. For r_2, we consider $\mathcal{P}_2 = (\mathcal{O}_2, \Gamma_2)$, $\mathcal{O}_2 = \{\text{drink(red-wine)} \succ \text{drink(white-wine)} \succ \text{drink(beer)}\}$, and $\Gamma_2 = \{\text{main-food(beef)}\}$. Furthermore, r_3 is specified as $\mathcal{P}_3 = (\mathcal{O}_3, \Gamma_3)$, $\mathcal{O}_3 = \{\text{appetizer(pie)} \succ \text{appetizer(ice-cream)}\}$, and $\Gamma_3 = \{\text{drink(beer)}\}$. Finally, we have $\Pi = \{\mathcal{P}_1, \mathcal{P}_2, \mathcal{P}_3\}$ and $\mathcal{O}_{MP} = \emptyset$. In ASO, M_2 is preferred to M_3. Similarly, in $\mathcal{P}\text{-}\mathcal{MS}$, we have $M_1 \succeq_{\mathcal{MP}} M_2$.

Theorem 4. *Let (P_{gen}, P_{pref}) be an ASO program, and $\mathcal{MP}$ be the representation of P_{pref} in $\mathcal{P}\text{-}\mathcal{MS}$. Given two ASP models M_1 and M_2 in ASO, Assume structures $\mathcal{A}$ and $\mathcal{B}$ in $\mathcal{P}\text{-}\mathcal{MS}$, represent M_1 and M_2 respectively. If M_1 is preferred to M_2 with respect to P_{pref}, then we have $\mathcal{A} \succ_{\mathcal{MP}} \mathcal{B}$ in $\mathcal{P}\text{-}\mathcal{MS}$.*

Proof: Assume $\{m_1, ..., m_n\}$ are models of an ASO program $P = (P_{gen}, P_{pref})$. Let's consider each model $m_i = \{l_1, l_2, ..., l_{n_i}\}$, generated by P_{gen}, is a set of literals. Each m_i can be associated to a mathematical structure $\mathcal{A}^i = (\mathcal{I}, D)$ where $D = \{True, False\}$ and $\mathcal{I}(m_i) = True$. So, P_{gen} is a modular system (a set of structures). We show how a preference rule in ASO formalism can be translated into a preference in $\mathcal{P}\text{-}\mathcal{MS}$.

Consider a preference rule $r : C_1 > ... > C_k \leftarrow a_1, ..., a_n, not\, b_1, ..., not\, b_n$. T A partial structure $\mathcal{B}_i$ (or a set of partial structures in case of logical combination of literals) can represent each C_i. The body of the rule is a set of positive and negative literals that can be simply expressed by a partial structure as well. A preference in the form of $\mathcal{P} = (\Gamma, \mathcal{O})$ can assert a preference rule in ASO. The body of such rule is modelled by Γ and $\mathcal{O}$ represents the head of the rule. Each rule r_i can be represented by a preference $\mathcal{P}_i$. Therefore, a set of rules P_{pref} can be presented by meta-preference $\mathcal{MP}$. ∎

5 Conclusion and Future Work

We proposed an abstract framework for unifying preference languages in modular systems. We introduced the notion of preference-based modular systems ($\mathcal{P}\text{-}\mathcal{MS}$). We demonstrated that a system obtained through combination of some $\mathcal{P}\text{-}\mathcal{MS}$ is also a $\mathcal{P}\text{-}\mathcal{MS}$. We studied how preferences expressed in other languages (three languages as examples) can be translated to our framework. Examples included three common preference languages: cp-nets, planning, and answer set optimization. Our future work will address expressivity and computational issues of the framework. We will continue our study of practical aspects of our framework in AI applications, in particular, Business Processes that have complex modular structures and different users may communicate through different formal languages.

References

[1] Erman Acar, Camilo Thorne, and Heiner Stuckenschmidt. Towards decision making via expressive probabilistic ontologies. In *Algorithmic Decision Theory*, pages 52–68. Springer, 2015.

[2] Jorge A. Baier and Sheila A. McIlraith. Planning with preferences. *AI Magazine*, 29(4):25–36, 2008.

[3] Stephan Borzsony, Donald Kossmann, and Konrad Stocker. The skyline operator. In *Data Engineering, 2001. Proceedings. 17th International Conference on IEEE*, pages 421–430. IEEE, 2001.

[4] Craig Boutilier, Ronen Brafman, Carmel Domshlak, Holger Hoos, and David Poole. Cp-nets: A tool for representing and reasoning with conditional ceteris paribus preference statements. *J. Artif. Intell. Res.(JAIR)*, 21:135–191, 2004.

[5] Craig Boutilier, Ronen Brafman, Carmel Domshlak, Holger H. Hoos, and David Poole. Preference-based constrained optimization with cp-nets. *Computational Intelligence*, 20(2):137–157, 2004.

[6] Ronen Brafman and Carmel Domshlak. Preference handling-an introductory tutorial. *AI magazine*, 30(1):58, 2009.

[7] Ronen I. Brafman, Carmel Domshlak, and Solomon Eyal Shimony. On graphical modeling of preference and importance. *J. Artif. Intell. Res. (JAIR)*, 25:389–424, 2006.

[8] Gerhard Brewka, James P Delgrande, Javier Romero, and Torsten Schaub. asprin: Customizing answer set preferences without a headache. In *AAAI*, pages 1467–1474, 2015.

[9] Gerhard Brewka, Ilkka Niemela, and Miroslaw Truszczynski. Answer set optimization. In *IJCAI*, volume 3, pages 867–872, 2003.

[10] Gerhard Brewka, Miroslaw Truszczynski, and Stefan Woltran. Representing preferences among sets. In *AAAI*, 2010.

[11] James Delgrande, Torsten Schaub, Hans Tompits, and Kewen Wang. A classification and survey of preference handling approaches in nonmonotonic reasoning. *Computational Intelligence*, 20(2):308–334, 2004.

[12] James P. Delgrande, Torsten Schaub, and Hans Tompits. A framework for compiling preferences in logic programs. *TPLP*, 3(2):129–187, 2003.

[13] James P. Delgrande, Torsten Schaub, and Hans Tompits. A general framework for expressing preferences in causal reasoning and planning. *J. Log. Comput.*, 17(5):871–907, 2007.

[14] Alireza Ensan and Eugenia Ternovska. Modular systems with preferences. In *Proceedings of International Joint Conference on Artificial Intelligence (IJCAI2015). Buenos-Aires, Argentina: AAAI Press*, 2015.

[15] Wolfgang Faber, Miroslaw Truszczyński, and Stefan Woltran. Abstract preference frameworks: a unifying perspective on separability and strong equivalence. In *Twenty-Seventh AAAI Conference on Artificial Intelligence*, 2013.

[16] Alan M Frisch, Warwick Harvey, Chris Jefferson, Bernadette Martínez-Hernández, and Ian Miguel. Essence: A constraint language for specifying combinatorial problems. *Constraints*, 13(3):268–306, 2008.

[17] Martin Gebser, Benjamin Kaufmann, and Torsten Schaub. Conflict-driven answer set solving: From theory to practice. *Artificial Intelligence*, 187:52–89, 2012.

[18] Werner Kiessling. Foundations of preferences in database systems. In *Proceedings of the 28th international conference on Very Large Data Bases*, pages 311–322. VLDB Endowment, 2002.

[19] Stephen Cole Kleene. *Introduction to metamathematics*. North-Holland Publishing Company, 1952.

[20] David Mitchell and Eugenia Ternovska. Clause-learning for modular systems. In *Logic Programming and Nonmonotonic Reasoning*, pages 446–452. Springer, 2015.

[21] David G. Mitchell and Eugenia Ternovska. A framework for representing and solving NP search problems. In *Proceedings, The Twentieth National Conference on Artificial Intelligence and the Seventeenth Innovative Applications of Artificial Intelligence Conference, Pittsburgh, Pennsylvania, USA*, pages 430–435, 2005.

[22] David G Mitchell and Eugenia Ternovska. Expressive power and abstraction in essence. *Constraints*, 13(3):343–384, 2008.

[23] Kenneth Ross. On the adequacy of partial orders for preference composition. In *DBRank Workshop*, 2007.

[24] Ganesh Ram Santhanam, Samik Basu, and Vasant Honavar. Representing and reasoning with qualitative preferences for compositional systems. *Journal of Artificial Intelligence Research*, 42(1):211–274, 2011.

[25] Shirin Sohrabi, Jorge A Baier, and Sheila A McIlraith. Htn planning with preferences. In *Twenty-First International Joint Conference on Artificial Intelligence*, 2008.

[26] Tran Cao Son and Enrico Pontelli. Planning with preferences using logic programming. In *Logic Programming and Nonmonotonic Reasoning*, pages 247–260. Springer, 2004.

[27] Kostas Stefanidis, Georgia Koutrika, and Evaggelia Pitoura. A survey on representation composition and application of preferences in database systems. *ACM Transactions on Database Systems (TODS)*, 36:19–28, 2011.

[28] Shahab Tasharrofi and Eugenia Ternovska. A semantic account for modularity in multi-language modelling of search problems. In *Frontiers of Combining Systems, 8th International Symposium, FroCoS 2011, Germany*, pages 259–274, 2011.

[29] Shahab Tasharrofi, Xiongnan Wu, and Eugenia Ternovska. Solving modular model expansion tasks. *arXiv preprint arXiv:1109.0583*, 2011.

[30] Eugenia Ternovska. An algebra of combined constraint solving. In Georg Gottlob, Geoff Sutcliffe, and Andrei Voronkov, editors, *GCAI 2015. Global Conference on Artificial Intelligence*, volume 36 of *EPiC Series in Computing*, pages 275–295. EasyChair, 2015.

[31] Eugenia Ternovska. Static and dynamic views on the algebra of modular systems. In *16TH INTERNATIONAL WORKSHOP ON NON-MONOTONIC REASONING (NMR 2016) Cape Town, South Africa, April 22-24, 2016*, 2016.

[32] Nic Wilson. Consistency and constrained optimisation for conditional preferences. In *Proceedings of the 16th Eureopean Conference on Artificial Intelligence, ECAI'2004, Valencia, Spain*, pages 888–894, 2004.

[33] Nic Wilson. Preference inference based on lexicographic models. In *ECAI*, pages 921–926, 2014.

A Framework for Versatile Knowledge and Belief Management Operations in a Probabilistic Conditional Logic

CHRISTOPH BEIERLE
University of Hagen, Faculty of Mathematics and Computer Science, 58084 Hagen, Germany
`christoph.beierle@fernuni-hagen.de`

MARC FINTHAMMER
University of Hagen, Faculty of Mathematics and Computer Science, 58084 Hagen, Germany
`marc.finthammer@fernuni-hagen.de`

NICO POTYKA
University of Hagen, Faculty of Mathematics and Computer Science, 58084 Hagen, Germany
`nico.potyka@fernuni-hagen.de`

JULIAN VARGHESE
University of Münster, Institute of Medical Informatics, 48149 Münster, Germany
`julian.varghese@ukmuenster.de`

GABRIELE KERN-ISBERNER
TU Dortmund, Department of Computer Science, 44221 Dortmund, Germany
`gabriele.kern-isberner@cs.uni-dortmund.de`

Abstract

Intelligent agents equipped with epistemic capabilities are expected to carry out quite different knowledge and belief management tasks like answering queries, performing diagnosis and hypothetical reasoning, or revising and updating their own state of belief in the light of new information. In any realistic setting, such agents must also take vagueness and uncertainty into account. In this paper, we report on an approach that uses probabilistic conditional logic for modelling such an intelligent agent. We propose a simple, yet powerful agent model supporting versatile knowledge and belief management operations. The semantics of a knowledge base consisting of a set of probabilistic conditionals is obtained by employing the principle of maximum entropy. We give an overview of the MEcoRe system providing the core functionalities needed for realizing the agent model. In order to illustrate the use of MEcoRe's functionalities, we present a case-study of applying probabilistic logic to the analysis of clinical patient data in neurosurgery. Probabilistic conditionals are used to build a knowledge base for modelling and representing both clinical brain tumor data and expert knowledge of physicians working in this area.

1 Introduction

Using probabilities for expressing uncertainty on the one hand and exploiting the rich power and formal groundings of a logical language on the other hand has a long tradition in knowledge representation and Artificial Intelligence (e.g. [44, 51, 14, 18]). Various forms of conditional logics have been proposed and studied for expressing uncertain, but reasonable, plausible, possible, or probable relationships between the antecedent and the consequence of a conditional (e.g. [1, 45, 7, 12, 6]).

Probabilistic conditionals can conveniently be used to model uncertain rules like "*If symptoms S_1, S_2, and S_3 are present, then there is a probability of 70% that the patient has disease D.*", which occur frequently in the medical domain. An intelligent agent providing decision support for performing medical diagnosis and for choosing a therapy must be able to deal with pieces of knowledge expressed by such rules, requiring elaborate knowledge representation and reasoning facilities. For instance, in neurosurgery, such an agent should be able to answer diagnostic questions in the presence of evidential facts like "*Given the evidence that the patient has perceptual disturbances, suffers from unusual pain in the head and that there are symptoms for increased intracranial pressure, what is the probability that he has a cranial nerve tumor?*", and the agent should be able to perform hypothetical reasoning as in: "*There is evidence that the patient has perceptual disturbances and that there are symptoms for increased intracranial pressure. If we chose a surgery for therapy and if the correct diagnosis was gliobastoma, what would be the patient's chance to recover completely without any serious complications?*" Moreover, when the agent lives in an uncertain and dynamic environment, she has to adapt her epistemic state constantly to changes in the surrounding world and to react adequately to new demands (cf. [10], [24]).

In this paper, we investigate and use probabilistic conditional logic from three different points of view. As the first main contribution, we present a simple, yet powerful agent model that is based on probabilistic conditional logic. We develop a framework supporting versatile knowledge and belief management operations. The semantics of a knowledge base consisting of a set of probabilistic conditionals is obtained by employing the principle of maximum entropy [58, 46, 27, 30]. Using this principle, a given knowledge base is not required to fulfill any particular completeness or independence constraints; instead, any missing or unspecified knowledge can be inductively completed in an information-theoretically optimal way. This inductive completion will be demonstrated, and we will show how answering queries and performing diagnosis and hypothetical reasoning is done. A special focus will be put on change operations that have to be carried out when new information becomes available. We distinguish between revision and update, and elaborate in detail iterative belief changes.

The second contribution of this paper is to present an overview of the MEcoRe system. MEcoRe realizes and implements the agent model and its belief management operations developed for probabilistic conditional logic under maximum entropy semantics.

In the third main contribution, we report on a case study on the application of probabilistic modelling and reasoning to clinical patient data in neurosurgery. A knowledge base BT representing and integrating both statistical frequencies of brain tumors reported in the literature as well as physicians' expert beliefs is developed and used to perform reasoning regarding the diagnosis of brain tumor types or the prognosis for patients (see [61, 60] for more information on the medical background). We show how reasoning with BT is done using MEcoRe, and we argue that the obtained results are well in accordance with a physician's point of view.

This article is a revised and largely extended version of [4]. The extensions cover in particular the presentation of the formal foundations of the underlying probabilistic conditional logic, a detailed elaboration of the evolution of an agent's epistemic state under iterated belief change operations, and concepts for implementing MEcoRe's belief management functionalities. The rest of this article is organized as follows. After recalling the required background of probabilistic conditional logic in Section 2, epistemic states and belief management operations for probabilistic logic under optimum entropy semantics are developed in Section 3. The MEcoRe system is presented in Section 4. In Section 5, the knowledge base BT is developed and used for illustrating MEcoRe's reasoning facilities. In Section 6, we conclude and point out further work.

2 Background: Probabilistic Conditional Logic

The basics of propositional logic with multi-valued variables and the language of conditionals are presented, followed by a specification of a probabilistic semantics of conditionals.

2.1 Propositional Logic with Multi-Valued Variables

We start with a propositional language $\mathcal{L}$ built up over a finite set Σ of possibly multi-valued propositional variables. Each variable $V \in \Sigma$ is associated with a set of values domain(V) called its domain. If domain(V) = $\{0, 1\}$, V is called a *Boolean* variable and a *multi-valued* variable otherwise. For instance, the variable warningSymptoms with domain $\{0, 1\}$ is a Boolean variable, while the variable therapy with domain $\{$conservative, surgery, none$\}$ is a multi-valued variable.

The set $\mathcal{L}$ of formulas over Σ is the smallest set containing the following elements:

1. If $V \in \Sigma$ and $v \in$ domain(V), then $(V = v) \in \mathcal{L}$, and $(V = v)$ is called an *atomic formula*.

2. If $F \in \mathcal{L}$, then $(\neg F) \in \mathcal{L}$.

3. If $F, G \in \mathcal{L}$, then $(F \wedge G) \in \mathcal{L}$.

The formulas of $\mathcal{L}$ will be denoted by uppercase Roman letters A, B, C and so on. In order to enhance readability, we will sometimes omit parentheses and the logical conjunction, writing AB instead of $A \wedge B$, and an overbar will indicate negation, i.e. $\overline{A}$ means $\neg A$. Moreover, if V is a Boolean variable, we will abbreviate the atomic formula $(V = 1)$ by just V, and the atomic formula $(V = 0)$ by $\neg V$ or by $\overline{V}$.

A possible world over $\mathcal{L}$ assigns to each variable from $\mathcal{L}$ a value from its domain, and Ω denotes the set of all possible worlds over $\mathcal{L}$. I.e., for $\Sigma = \{V_1, \ldots, V_n\}$, a possible world $\omega \in \Omega$ is of the form

$$\omega = (V_1 = v_1) \wedge \ldots \wedge (V_n = v_n)$$

where $v_i \in$ domain(V_i) for $i \in \{1, \ldots, n\}$. Thus, as usual for propositional logic, each possible world ω can be identified uniquely with an interpretation I_ω; this interpretation is given by

$$I_\omega : \Sigma \to \bigcup_{V \in \Sigma} \text{domain}(V)$$

with $I(V) \in$ domain(V) for all $V \in \Sigma$ such that $I(V) = v$ iff $(V = v)$ appears in ω.

The satisfaction relation $\models$ between possible worlds and formulas is defined recursively, for every $\omega \in \Omega$, as follows:

1. If $(V\!=\!v)$ is an atomic formula, then $\omega \models (V\!=\!v)$ if and only if $I_\omega(V) = v$.

2. If $F \in \mathcal{L}$, then $\omega \models \neg F$ if and only if $\omega \models F$ does not hold.

3. If $F, G \in \mathcal{L}$, then $\omega \models F \wedge G$, if and only if $\omega \models F$ and $\omega \models G$.

Note that for the case where all variables in Σ are Boolean variables, we obtain standard propositional logic. For instance, for $\Sigma = \{a, b\}$ with Boolean variables a, b, the set of possible worlds (or interpretations) is given by $\Omega = \{ab, a\bar{b}, \bar{a}b, \bar{a}\bar{b}\}$.

2.2 Conditionals

By introducing a new binary operator $|$, we can define the set of all (unquantified) *conditionals* (or *rules*) over $\mathcal{L}$ as

$$(\mathcal{L} \mid \mathcal{L}) = \{(B|A) \mid A, B \in \mathcal{L}\}.$$

A conditional of the form $(B|\top)$ with a tautological antecedent $\top$ is also called a *fact* and can simply be written as (B). Intuitively, $(B|A)$ expresses "*if A then B*" and establishes a plausible (or probable, possible, etc.) connection between the *antecedent* A and the *consequent* B. Here, we are interested in the set of all *probabilistic conditionals* (or *probabilistic rules*)

$$(\mathcal{L} \mid \mathcal{L})^{prop} = \{(B|A)[x] \mid A, B \in \mathcal{L}, x \in [0, 1]\},$$

which contains conditionals enriched with a probability that reflects the likelihood of B given A. Thus, a conditional $(B|A)[x]$ is read as "*if A holds, then B holds with probability x*". The powerset of probabilistic conditionals is denoted by

$$\mathcal{COND} = \{\mathcal{R} \mid \mathcal{R} \subseteq (\mathcal{L} \mid \mathcal{L})^{prop}\}.$$

Syntactically, our probabilistic conditional language is closely related to probabilistic logic programming as studied in [37, 30] where intervals of lower and upper probabilities are allowed, while in this paper we only consider point probabilities.

2.3 Probabilistic Semantics

To give appropriate semantics to conditionals, they are usually considered within richer structures such as *epistemic states*. Besides certain (logical) knowledge, epistemic states also allow the representation of e.g. preferences, beliefs, assumptions of an intelligent agent. Basically, an epistemic state allows one to compare formulas or worlds with respect to plausibility, possibility, necessity, probability etc.

In a quantitative framework, popular representations of epistemic states are provided by *probability functions* (or *probability distributions*). Given the underlying propositional language $\mathcal{L}$ and its induced set of possible worlds Ω, a corresponding probability distribution

$$P : \Omega \to [0,1]$$

with $\sum_{\omega \in \Omega} P(\omega) = 1$ assigns a probability to each possible world. The function P is extended to arbitrary formulas, denoted by $P : \mathcal{L} \to [0,1]$, by defining

$$P(A) = \sum_{\omega \in \Omega, \omega \models A} P(\omega)$$

for every $A \in \mathcal{L}$.

Thus, in this setting, the set of *epistemic states* we will consider is

$$EpState = \{P \mid P : \Omega \to [0,1] \text{ is a probability function}\}.$$

In this structure, conditionals are interpreted via conditional probability. The probability of a conditional $(B|A) \in (\mathcal{L} \mid \mathcal{L})$ with $P(A) > 0$ is defined as $P(B|A) = P(AB)/P(A)$, the corresponding conditional probability. So the satisfaction relation

$$\models_c \; \subseteq \; EpState \times (\mathcal{L} \mid \mathcal{L})^{prop}$$

of probabilistic conditional logic is defined by

$$P \models_c (B|A)[x] \quad \text{iff} \quad P(B|A) = x.$$

As usual, the satisfaction relation $\models_c$ is extended to a set $\mathcal{R} \in \mathcal{COND}$ of conditionals by defining $P \models_c \mathcal{R}$ iff $P \models_c (B|A)[x]$ for all $(B|A)[x] \in \mathcal{R}$. A set $\mathcal{R}$ of conditionals is consistent iff there is a probability distribtion P such that $P \models_c \mathcal{R}$; otherwise, $\mathcal{R}$ is inconsistent.

3 Epistemic States and Belief Management Operations for Probabilistic Logic under Optimum Entropy

Using probability distributions for modelling the epistemic state of an intelligent agent allows to take the uncertainty of knowledge and belief into account. In this section, we consider the main belief management tasks of such an intelligent agent like abductive reasoning or revising or updating her own state of belief, and model them in the setting of probabilistic conditional logic. In [5], related knowledge and belief operations for a conceptual agent model in a setting based on qualitative default rules are given, using the general approach to belief change developed in [27, 29] where belief change is considered in a very general form by revising epistemic states by sets of conditionals.

3.1 Initialization

Initially, an epistemic state has to be built up on the basis of which the agent can start her computations. If no knowledge is at hand, the epistemic state should reflect complete ignorance. In our probabilistic setting, this corresponds to the uniform distribution where the same probability is assigned to each possible world. If, however, a set of probabilistic rules is at hand to describe our problem domain, we have to determine an epistemic state that represents our prior knowledge appropriately. To this end, we assume an inductive representation method to establish the desired connection between sets of sentences and epistemic states. Whereas generally, a set $\mathcal{R}$ of sentences allows a (possibly large) set of models (or epistemic states), in an inductive formalism we have a function

$$inductive : \mathcal{COND} \to EpState$$

such that $inductive(\mathcal{R})$ selects a unique, "best" epistemic state from all those states satisfying $\mathcal{R}$.

In a probabilistic framework, a probability distribution is a suitable representation of an epistemic state. Given a set $\mathcal{R}$ of probabilistic conditionals, the *principle of maximum entropy* states to select the distribution $P^* = MaxEnt(\mathcal{R})$ that satisfies all conditionals in $\mathcal{R}$ and maximizes entropy among all distributions that do so. More formally, $MaxEnt(\mathcal{R})$ is the solution to the maximization problem

$$\arg\max_{P' \models \mathcal{R}} H(P') \quad (1)$$

with the entropy $H(P')$ of a distribution P' being defined by

$$: H(P') = -\sum_{\omega} P'(\omega) \log P'(\omega) \quad (2)$$

If $\mathcal{R}$ is consistent, there exists indeed a unique solution P^*. This follows from the fact that the set of all distributions that satisfy $\mathcal{R}$ form a compact and convex set and the entropy is a continuous and strictly concave function, see [46] for the details.

The rationale behind this is that $MaxEnt(\mathcal{R})$ represents the knowledge given by $\mathcal{R}$ most faithfully, i.e. without adding information unnecessarily (cf. [23, 49]). The principle of maximum entropy has also been justified by properties of commonsense reasoning [47] and by conditional-logical considerations [26, 25]. In contrast to Bayes nets [51], which are based on assumptions about conditional independencies and require certain conditional probabilities to be given, the principle of maximum entropy completes the actually available knowledge $\mathcal{R}$ in an information-theoretically optimal way. Thus, aside from consistency, the available knowledge $\mathcal{R}$ does not have to comply with any additional requirements, thereby avoiding any unwanted bias in the knowledge modelling process.

We will illustrate the maximum entropy method by a small example.

Example 1 (Running example; Initialization). *Consider the three propositional Boolean variables:*

$$\begin{aligned} s : &\quad \textit{being a student} \\ y : &\quad \textit{being young} \\ u : &\quad \textit{being unmarried} \end{aligned}$$

Students and unmarried people are mostly young. This commonsense knowledge an agent may have can be expressed probabilistically e.g. by the set

$$\mathcal{R}_1 = \{\ (y|s)[0.8],\ (y|u)[0.7]\ \}$$

of conditionals. The MaxEnt-representation $P_1^* = MaxEnt(\mathcal{R}_1)$ over the set of possible worlds $\Omega = \{syu, sy\bar{u}, s\bar{y}u, \ldots \bar{s}\,\bar{y}\,\bar{u}\}$ is computed by MECoRe *by employing an iterative algorithm (which we will consider more closely in Section 4.1). The result (modulo rounding) is:*

ω	$P_1^*(\omega)$	ω	$P_1^*(\omega)$	ω	$P_1^*(\omega)$	ω	$P_1^*(\omega)$
syu	0.1950	$sy\bar{u}$	0.1758	$s\bar{y}u$	0.0408	$s\bar{y}\bar{u}$	0.0519
$\bar{s}yu$	0.1528	$\bar{s}y\bar{u}$	0.1378	$\bar{s}\bar{y}u$	0.1081	$\bar{s}\bar{y}\bar{u}$	0.1378

3.2 Querying an Epistemic State

Querying an agent about her beliefs amounts to posing a set of unquantified sentences and asking for the corresponding degrees of belief with respect to her current epistemic state. In our probabilistic framework, a query is an unquantified conditional $(B|A)$ and we are interested in a probability x such that we have $P^* \models_c (B|A)[x]$ for our epistemic state P^*.

Example 2 (Query). *Suppose the current epistemic state is $P_1^* = MaxEnt(\mathcal{R}_1)$ from Example 1, and our question is:*

"What is the probability that unmarried students are young?"

So, the query is expressed by the unquantified conditional $(y|s \wedge u)$. *Then* MECoRe *returns:*

$$(y|s \wedge u)[0.8270]$$

That is, unmarried students are supposed to be young with probability 0.8270.

3.3 New Information and Belief Change

Belief revision, the theory of dynamics of knowledge, has been mainly concerned with propositional beliefs for a long time. The most basic approach here is the *AGM-theory* presented in the seminal paper [2] as a set of postulates outlining appropriate revision mechanisms in a propositional logical environment. This framework has been widened by Darwiche and Pearl [10] for (qualitative) epistemic states and conditional beliefs in order to be able to do revision iteratively. An even more general approach, unifying revision methods for quantitative and qualitative representations of epistemic states, is described in [27]. The crucial meaning of conditionals as *revision policies* for belief revision processes is made clear by the so-called *Ramsey test*, according to which a conditional $(B|A)$ is accepted in an epistemic state Ψ, iff revising Ψ by A yields belief in B: $\Psi \models (B|A)$ iff $\Psi * A \models B$ where $*$ is a belief revision operator (see e.g. [17]).

Beside belief revision, also belief updating [24] has been studied for belief change. Generally, it is understood that *revision* takes place when new information about a static world arrives, whereas *updating* tries to incorporate new information about a (possibly) evolving, changing world. Further belief change operators are *expansion*, *focusing*, *contraction*, and *erasure* (cf. [17, 13, 24]). However, the techniques that have been proposed for such change operations are mostly limited to the traditional AGM framework for change, i.e., to propositional logic. In this paper, we rely upon the more general approach to epistemic change developed in [27] where belief change is considered in a very general and advanced form: Epistemic states are revised by sets of conditionals – this exceeds by far the classical AGM-theory and related approaches which only deal with pieces, or sets of propositional beliefs. Nevertheless, connections to belief change theory can be found when looking into works on iterated belief change [10] that extends AGM to the epistemic level. This has been elaborated in detail e.g. in [28]. Therefore, the belief change operations to be used in the following share the underlying intention and rationale with the propositional change operations of the respective same name. In particular, we stick to the view that revision is a change operation within a static context, while updating may take changes in the world, or context, respectively, into account.

In the following, we present in particular how the core belief change functionalities of update and revision are realized in our setting.

3.3.1 Belief Change Operator *MinCEnt*

In the probabilistic framework, a powerful operator to change probability distributions by sets of probabilistic conditionals is provided by the *principle of minimum cross-entropy* which generalizes the principle of maximum entropy in the sense of (1): Given a (prior) distribution P and a set $\mathcal{R}$ of probabilistic conditionals, the *MinCEnt-distribution* $P^* = MinCEnt(P, \mathcal{R})$ obtained by the function

$$MinCEnt : EpState \times \mathcal{COND} \to EpState \qquad (3)$$

is the distribution that satisfies all constraints in $\mathcal{R}$ and has minimal cross-entropy (also called Kullback-Leibler divergence) H_{ce} with respect to P, i.e. P^* solves the minimization problem

$$\arg\min_{P' \models \mathcal{R}} H_{ce}(P', P) \qquad (4)$$

with:

$$H_{ce}(P', P) = \sum_{\omega} P'(\omega) \log \frac{P'(\omega)}{P(\omega)} \qquad (5)$$

If $\mathcal{R}$ is basically compatible with P (i.e. *P-consistent*, cf. [27]), then again there exists a unique solution P^* (for further information and lots of examples, see [9, 49, 27]). The cross-entropy between two distributions can be taken as a directed (i.e. asymmetric) information distance [57] between these two distributions. Following the principle of minimum cross-entropy means to modify the prior epistemic state P in such a way as to obtain a new distribution P^* which satisfies all conditionals in $\mathcal{R}$ and is as close to P as possible. So, the *MinCEnt*-principle yields a probabilistic belief change operator, associating to each probability distribution P and each P-consistent set $\mathcal{R}$ of probabilistic conditionals the changed distribution

$$P^* = MinCEnt(P, \mathcal{R})$$

in which $\mathcal{R}$ holds. Similar to the *MaxEnt*-principle, the *MinCEnt*-principle can be axiomatized from first principles [26] and shows an excellent handling of conditional relationships [28] under change operations. Indeed, it generalizes Bayesian conditionalization: If $\mathcal{R}$ consists of just one piece of sure information, i.e., $\mathcal{R} = \{A[1]\}$, then $MinCEnt(P, \mathcal{R}) = P(\cdot | A)$. However, please note that in general,

$$MinCEnt(P, \mathcal{R} \cup \mathcal{S}) \neq MinCEnt(MinCEnt(P, \mathcal{R}), \mathcal{S}).$$

Observing (and indeed relying upon) this subtle difference will be crucial in the rest of this section.

3.3.2 Belief Bases

Both probabilistic revision and update operations can be defined by means of the belief change operator *MinCEnt*. For the distinction between revision and update, we have to distinguish whether the new knowledge to be taken into account reflects new information about an unchanged, static world, or whether it is due to a change of the world. Towards this end, we introduce the notion of a *belief base* which is a pair

$$\langle P, \mathcal{R} \rangle$$

consisting of a probability distribution P and a knowledge base $\mathcal{R}$. Here, P is understood as a kind of background knowledge that is used for adapting to the rules explicitly stated in $\mathcal{R}$. For instance, P can reflect the generic knowledge that an employee has about how usually companies work and people in companies behave, while $\mathcal{R}$ may contain information about the structure and work flows of a specific company in which the employee has just started to work. The employee is still able to follow general rules of business life (encoded in P), but she can also change, modify, or adapt such rules according to $\mathcal{R}$. In particular, more specific information in $\mathcal{R}$ can override general knowledge in P in case of conflict. However, if there is no conflict between knowledge in P established from experience and new information in $\mathcal{R}$, this established knowledge is still available to the employee.

The set of belief bases is denoted by *BBase*, and the function

$$ES : BBase \to EpState \qquad (6)$$
$$ES(\langle P, \mathcal{R} \rangle) \mapsto MinCEnt(P, \mathcal{R})$$

yields the *epistemic state induced by a belief base*. Thus, for the belief base $\langle P, \mathcal{R} \rangle$ the induced epistemic state is

$$P^* = ES(\langle P, \mathcal{R} \rangle) = MinCEnt(P, \mathcal{R})$$

i.e., the probability distribution that is obtained from P and $\mathcal{R}$ by applying our belief change operator. In this way, we distinguish between the explicit beliefs specified by $\mathcal{R}$, and the implicit knowledge in P^* that is derived from P and $\mathcal{R}$.

Example 3 (Belief base). *Let us reconsider Example 1 with our new concepts. Our initial belief base is $B_1 = \langle P_0, \mathcal{R}_1 \rangle$, where P_0 is the uniform distribution over the possible worlds in Example 1 and $\mathcal{R}_1$ is the knowledge base from Example 1. Note that maximizing entropy corresponds to minimizing relative entropy to the uniform distribution. Therefore, the epistemic state induced by B_1 is the probability distribution P_1^* from Example 1 since $MinCEnt(P_0, \mathcal{R}_1) = P_1^*$.*

3.3.3 Revision

Our revision operator deals with new knowledge in a static world, where the belief state may change but the explicit knowledge remains valid. Therefore, in contrast to update operations, revision is performed under the assumption that the new knowledge is consistent with the old explicit knowledge and extends this knowledge. Former explicit knowledge remains valid, even though the corresponding epistemic state usually changes. The *revision operator* ∘ taking a belief base and a set of conditionals as input and yielding a new belief base is defined by:

$$\_ \circ \_ : BBase \times \mathcal{COND} \to BBase \qquad (7)$$
$$\langle P, \mathcal{R} \rangle \circ \mathcal{R}' \mapsto \langle P, \mathcal{R} \cup \mathcal{R}' \rangle$$

Thus, the epistemic state induced after a revision operation is given by:

$$ES(\langle P, \mathcal{R} \rangle \circ \mathcal{R}') = MinCEnt(P, \mathcal{R} \cup \mathcal{R}') \qquad (8)$$

Note that the component P of the belief base is not changed by a revision, but that the former explicit knowledge $\mathcal{R}$ is preserved. This also illustrates the similarities between revision and expansion which have also been studied in AGM theory [2]. But note that while revision is realized by an expansion on the level of belief bases, it yields a true, non-trivial revision on the level of epistemic states.

Example 4 (Revision). *Consider again the belief base $B_1 = \langle P_0, \mathcal{R}_1 \rangle$ from Example 3. Suppose that we learn that in our population there are 40% students. We revise our belief base appropriately and get the new belief base*

$$B_2 = \langle P_0, \mathcal{R}_1 \cup \{(s)[0.4]\} \rangle.$$

The new induced epistemic state

$$P_2^* = MinCEnt(P_0, \mathcal{R}_1 \cup \{(s)[0.4]\})$$

as determined by MEcoRe *is:*

ω	$P_2^*(\omega)$	ω	$P_2^*(\omega)$	ω	$P_2^*(\omega)$	ω	$P_2^*(\omega)$
syu	0.1703	$sy\overline{u}$	0.1496	$s\overline{y}u$	0.0340	$s\overline{y}\,\overline{u}$	0.0459
$\overline{s}yu$	0.1760	$\overline{s}y\overline{u}$	0.1547	$\overline{s}\,\overline{y}u$	0.1144	$\overline{s}\,\overline{y}\,\overline{u}$	0.1547

While the epistemic state P_2^* is different from the epistemic state P_1^* from Examples 1 and 3, also P_2^* satisfies all conditionals in $\mathcal{R}_1$, i.e., $P_2^* \models_c (y|s)[0.8]$ and $P_2^* \models_c (y|u)[0.7]$.

3.3.4 Update

Intuitively, our update operation deals with new beliefs in a dynamic world (or context) where the background knowledge may change. So, update does not presuppose consistency between old and new explicit knowledge, and former explicit knowledge does not necessarily remain valid but becomes part of the background knowledge. The *update operator* • taking a belief base and a set of conditionals as input and yielding a new belief base is defined by:

$$\_ \bullet \_ : BBase \times \mathcal{COND} \to BBase \qquad (9)$$
$$\langle P, \mathcal{R} \rangle \bullet \mathcal{R}' \mapsto \langle MinCEnt(P, \mathcal{R}), \mathcal{R}' \rangle$$

So, the new background knowledge after the update is $MinCEnt(P, \mathcal{R})$, and the new explicit knowledge is $\mathcal{R}'$. Thus, the epistemic state induced after an update operation is given by

$$ES(\langle P, \mathcal{R} \rangle \bullet \mathcal{R}') = MinCEnt(MinCEnt(P, \mathcal{R}), \mathcal{R}') \qquad (10)$$

which is known to be different from $MinCEnt(P, \mathcal{R} \cup \mathcal{R}')$ (cf. Equation 8).

When comparing (7) and (8) with (9) and (10), the methodological differences between revision and update operations in our framework become clear: Whereas former explicit knowledge is preserved under revision, it becomes part of the background knowledge under updating.

Example 5 (Update). *Consider again the belief base $B_2 = \langle P_0, \mathcal{R}_2 \rangle$ with*

$$\mathcal{R}_2 = \mathcal{R}_1 \cup \{(s)[0.4]\} = \{(y|s)[0.8], (y|u)[0.7], s[0.4]\}$$

from Example 4. Suppose that some time later, the relationship between students and young people has changed in the population, so that students are young with a probability of 0.9. In order to incorporate this new knowledge $(y|s)[0.9]$ which is in conflict with $\mathcal{R}_2$, the agent applies an updating operation to modify the belief base B_2 appropriately. The new belief base is

$$B_3 = \langle MinCEnt(P_0, \mathcal{R}_2), \{(y|s)[0.9]\} \rangle.$$

The new epistemic state induced by B_3 is

$$P_3^* = MinCEnt(MinCEnt(P_0, \mathcal{R}_2), \{(y|s)[0.9]\})$$
$$= MinCEnt(P_2^*, \{(y|s)[0.9]\})$$

where P_2^* is as in Example 4. P_3^* as determined by MEcoRe *is:*

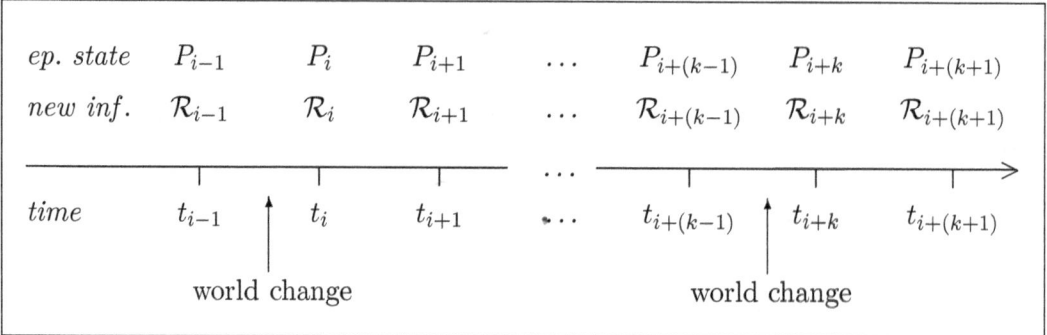

Figure 1: Time scale for iterated belief change.

ω	$P_3^*(\omega)$	ω	$P_3^*(\omega)$	ω	$P_3^*(\omega)$	ω	$P_3^*(\omega)$
syu	0.1874	$sy\bar{u}$	0.1646	$s\bar{y}u$	0.0166	$s\bar{y}\bar{u}$	0.0224
$\bar{s}yu$	0.1786	$\bar{s}y\bar{u}$	0.1569	$\bar{s}\bar{y}u$	0.1161	$\bar{s}\bar{y}\bar{u}$	0.1569

It is easily checked that indeed, $P_3^*(y|s) = 0.9$ (taking rounding into account), whereas $P_2^*(y|s) = 0.8$. With the update operation, also the probabilities of the other conditionals in $\mathcal{R}_2$ have changed, in an entropy-optimal way by minimizing the cross-entropy from P_2^* to the epistemic state P_3^* satisfying the new knowledge $(y|s)[0.9]$; for instance, we have $P_2^*(y|u) = 0.7$, and $P_3^*(y|u) = 0.73$. Thus, updating the agent's belief base with an increased probability for the conditional knowledge that students are young also slightly increases the agent's belief that unmarried people are young.

3.3.5 Iterated Belief Changes

We will now demonstrate how iterated belief change is realized by means of the concepts introduced above. Our illustration focuses on how the epistemic state of an agent changes over time in a sequence of revision and update steps.

When considering the process of iterated belief change, we may assume a discrete time scale as sketched in Figure 1 (cf. [5]). For any j, P_j is the agent's epistemic state at time t_j, and $\mathcal{R}_j$ is the new information given to the agent at time t_j. Furthermore, for any j, the agent has to adapt her own epistemic state P_j in the light of the new information $\mathcal{R}_j$, yielding P_{j+1}. Now suppose that at times t_i and t_{i+k} the agent receives the information that the world has changed (i.e., a world change occurred between t_{i-1} and t_i and between $t_{i+(k-1)}$ and t_{i+k}). Thus, the changes from P_{i-1} to P_i and from $P_{i+(k-1)}$ to P_{i+k} are done by an update operation. Suppose further, that the world remains static in between, implying that all belief changes from P_i up to $P_{i+(k-1)}$ are revisions. Furthermore, let the world remain static also after t_{i+k}, i.e., the belief change from P_{i+k} to $P_{i+(k+1)}$ is also achieved by a revision.

Now using the binary belief change operator *MinCEnt* as described in Section 3.3.1, we can precisely describe how the agent's epistemic states

$$P_i, P_{i+1}, \ldots, P_{i+(k-1)}, P_{i+k}, P_{i+(k+1)}$$

are obtained. Denoting *MinCEnt* by the infix operator $\diamond_{me}$, we have the following situation:

$$
\begin{aligned}
\text{update}: & \quad P_i & = & \quad P_{i-1} \diamond_{me} (\mathcal{R}_{i-1}) \\
\text{revision}: & \quad P_{i+1} & = & \quad P_{i-1} \diamond_{me} (\mathcal{R}_{i-1} \cup \mathcal{R}_i) \\
\text{revision}: & \quad P_{i+2} & = & \quad P_{i-1} \diamond_{me} (\mathcal{R}_{i-1} \cup \mathcal{R}_i \cup \mathcal{R}_{i+1}) \\
& \quad \ldots \\
\text{revision}: & \quad P_{i+(k-1)} & = & \quad P_{i-1} \diamond_{me} (\mathcal{R}_{i-1} \cup \mathcal{R}_i \cup \ldots \cup \mathcal{R}_{i+(k-2)}) \\
\text{update}: & \quad P_{i+k} & = & \quad P_{i+(k-1)} \diamond_{me} (\mathcal{R}_{i+(k-1)}) \\
\text{revision}: & \quad P_{i+(k+1)} & = & \quad P_{i+(k-1)} \diamond_{me} (\mathcal{R}_{i+(k-1)} \cup \mathcal{R}_{i+k})
\end{aligned}
$$

For the realization of such an agent with evolving epistemic state it suffices to store and modify a single belief base $\langle P, \mathcal{R} \rangle$ where:

- P denotes the epistemic state of the agent before the last update took place, and

- $\mathcal{R}$ is the union of all probablitistic conditionals that the agent received as new information and that have been used to revise her epistemic state since the last update occurred.

In Section 4, we will describe how this approach is implemented in the MECoRe system.

3.4 Diagnosis

A very common operation in knowledge-based systems is to come up with a diagnosis for a given case. Given some case-specific *evidence* E (formally, a set of quantified facts), diagnosis assigns degrees of belief to the atomic propositions D to be *diagnosed* (formally, $D = \{d_1, \ldots d_n\}$ is a set of unquantified atomic propositions). Thus, making a diagnosis in the light of some given evidence corresponds to determine what is believed in the state obtained by focusing the current state P on the given evidence, i.e. querying the epistemic state $MinCEnt(P, E)$ with respect to D. Thus, here focusing corresponds to conditioning P with respect to the given evidence E since we are looking for the probabilities x_i such that

$$MinCEnt(P, E)(d_i) = x_i$$

holds.

Example 6. *Let P_1^* be the epistemic state from Example 3. If there is now certain evidence for being a student and being unmarried, i.e., $E = \{(s \wedge u)[1.0]\}$, and we ask for the degree of belief of being young, i.e., $D = \{y\}$, MEcore computes $y[0.8270]$ since*

$$MinCEnt(P_1^*, \{(s \wedge u)[1.0]\})(y) = 0.8270.$$

Thus, if there is certain evidence for being an unmarried student, then the degree of belief for being young is 0.8270.

While the given evidence in Example 6 is certain since its probability is 1.0, of course also uncertain evidence with any probability can be handled in the same way for diagnosis.

3.5 What-If-Analysis: Hypothetical Reasoning

Hypothetical reasoning asks for the degree of belief of complex relationships (goals) under some hypothetical assumptions. This is useful, e.g., to exploit in advance the benefits of some expensive or intricate medical investigations. Note that whereas in the diagnostic case both evidence E and diagnoses D are just simple propositions, in hypothetical reasoning both the *assumptions* A (formally, a set of quantified conditionals) as well as the *goals* G (formally, $G = \{(L_1|K_1), \ldots (L_n|K_n)\}$ is a set of unquantified conditionals) may be sets of full conditionals. However, since the underlying powerful *MinCEnt*-change operator can modify epistemic states by arbitrary sets of conditionals, our framework can handle hypothetical what-if-analysis structurally analogously to the diagnostic case, i.e. by conditioning the current epistemic state P with respect to the assumptions A and querying the resulting state with respect to G. Thus, we are looking for the probabilities x_i such that

$$MinCEnt(P, A)(L_i|K_i) = x_i$$

holds. Note that since this is hypothetical reasoning, the agent's current epistemic state remains unchanged.

Example 7. *Given P_2^* from Example 4 as present epistemic state, a hypothetical reasoning question is given by:*

> *"What would be the probability of being young under the condition of being unmarried, provided that the probability of a student being young changed to 0.9?"*

So we have the goal $G = \{(y|u)\}$ and the assumption $A = \{(y|s)[0.9]\}$. Then MEcore's answer is $(y|u)[0.73]$ since

$$MinCEnt(P_2^*, \{(y|s)[0.9]\})(y|u) = 0.73.$$

Note that this probability is the probability given by P_3^* from Example 5 since

$$MinCEnt(P_2^*, \{(y|s)[0.9]\}) = P_3^*,$$

but unlike the update operation in Example 5, hypothetical reasoning neither changes the agent's belief base nor her epistemic state.

4 The MEcoRe system

Besides providing the core functionalities needed for probabilistic reasoning at optimum entropy, the main objective of MEcoRe is to support advanced belief management operations like revision, update, diagnosis, or what-if-analysis in a most flexible and easily extendible way[1]. MEcoRe is implemented in Java and provides several implementations for computing $MinCEnt(P, \mathcal{R})$.

4.1 Computation of *MinCEnt*

As elaborated in Section 3, the change function *MinCEnt* lies at the core of almost all knowledge and belief management operations in MEcoRe. Computing $MinCEnt(P, \mathcal{R})$ requires to solve the numerical optimization problem given by Expression (4) in Section 3.3.1. MEcoRe provides two alternative algorithms to determine the probability distribution $MinCEnt(P, \mathcal{R})$ induced by a given belief base $\langle P, \mathcal{R} \rangle$, and one can also choose optimized versions of these algorithms to exploit special structures of knowledge bases.

The first basic algorithm is an implementation of Csiszar's *Iterative I-Projections* algorithm [9]. An I-projection is a projection that is defined by information-theoretic means as explained in the following. The algorithm starts with P and then iterates over the conditionals in $\mathcal{R}$. In each iteration, a conditional $r \in \mathcal{R}$ is selected and the current distribution P is *I-projected* on the set of probability distributions that satisfy r. That is, the next distribution P' is the distribution $MinCEnt(P, \{r\})$ that satisfies r and minimizes the cross-entropy to P. This process converges to $MinCEnt(P, \mathcal{R})$ if $\mathcal{R}$ is consistent [9]. Note, however, that in each iteration $MinCEnt(P, \{r\})$ has to be computed for some $r \in \mathcal{R}$ and it is not at all clear how this is easier than computing $MinCEnt(P, \mathcal{R})$. Indeed, the main challenge in implementing Iterative I-Projections is to find an efficient way to compute $MinCEnt(P, \{r\})$. Fortunately, in our framework, $MinCEnt(P, \{r\})$ can be obtained by a simple update formula (see [41], Lemma 4.2).

---

[1] MEcoRe is available at:
http://sourceforge.net/p/kreator-ide/code/HEAD/tree/Software/MECore/

MECoRe's second basic algorithm is based on *Limited Memory BFGS (L-BFGS)* [35]. Roughly speaking, BFGS is a Quasi-Newton method which approximates Newton's method to find local optima of functions. Newton's method is a line-search method like gradient descent, but it takes second-order information from the Hessian matrix into account and in this way provides better convergence guarantess. However, computing the Hessian matrix is too expensive in many applications, therefore Quasi-Newton methods use an approximation. L-BFGS is a memory-optimized version that avoids storing the whole matrix and turned out to be very efficient for probabilistic reasoning problems [40, 42]. In our framework, Iterative I-Projections is often the better choice for small or loosely correlated knowledge bases, whereas L-BFGS becomes preferable for more complex knowledge bases (see [53] for some benchmarks).

Both algorithms can be optimized if the knowledge base features special structure. First, deterministic conditionals with probability 0 or 1 let worlds become meaningless by enforcing zero probabilities. These worlds can be deleted in a preprocessing step to decrease the size of the probability tables. Second, worlds are often *indifferent* with respect to the ME-optimal distribution and can be combined in an equivalence class to further reduce the number of worlds [16, 30, 15].

4.2 Performing Revision More Efficiently

Once a belief base $\langle P, \mathcal{R} \rangle$ is defined or modified in MECoRe, the current epistemic state

$$P^*_{\text{cur}} := ES(\langle P, \mathcal{R} \rangle) = MinCEnt(P, \mathcal{R}) \qquad (11)$$

induced by $\langle P, \mathcal{R} \rangle$ is computed immediately. Since determining $MinCEnt(P, \mathcal{R})$ by one of the available algorithms (see Section 4.1) may be computationally expensive, the epistemic state P^*_{cur} induced by $\langle P, \mathcal{R} \rangle$ is only determined once and stored afterwards for later use, e.g., for answering queries. Therefore, for a pair $\langle P, \mathcal{R} \rangle$ representing a belief base, actually the triple $(P, \mathcal{R}, P^*_{\text{cur}})$ is stored in MECoRe.

If the current belief base $\langle P, \mathcal{R} \rangle$ is revised by a set of conditionals $\mathcal{R}'$, i.e. when performing the revision

$$\langle P, \mathcal{R} \rangle \circ \mathcal{R}' = \langle P, \mathcal{R} \cup \mathcal{R}' \rangle, \qquad (12)$$

then MECoRe has to compute the revised epistemic state

$$P^*_{\text{rev}} := ES(\langle P, \mathcal{R} \cup \mathcal{R}' \rangle) = MinCEnt(P, \mathcal{R} \cup \mathcal{R}') \qquad (13)$$

induced by the revised belief base $\langle P, \mathcal{R} \cup \mathcal{R}' \rangle$. However, MEcoRe also allows to compute P^*_{rev} in a more efficient way by exploiting the fact that the following equation holds for *MinCEnt* (as shown in [27]):

$$MinCEnt(P, \mathcal{R} \cup \mathcal{R}') = MinCEnt(\underbrace{MinCEnt(P, \mathcal{R})}_{P^*_{cur}}, \mathcal{R} \cup \mathcal{R}') \tag{14}$$

That is, MEcoRe can directly consider the current epistemic state P^*_{cur} and compute

$$P^*_{rev} = MinCEnt(P^*_{cur}, \mathcal{R} \cup \mathcal{R}'). \tag{15}$$

In many cases, considering P^*_{cur} (as in (15)) instead of P (as in (13)) when computing P^*_{rev} can be more efficient, since P^*_{cur} already satisfies all of the conditionals in $\mathcal{R}$ (cf. (11)). That is, the actual computation of *MinCEnt* can often be performed faster by starting from P^*_{cur}, since P^*_{cur} is already "closer" to P^*_{rev}.

4.3 The User Interface

MEcoRe can be controlled by a text command interface or by scripts, i.e. text files that allow the batch processing of command sequences. These scripts and the text interface use a programming language-like syntax that allows to define, manipulate and display variables, propositions, rule sets and epistemic states. The following sequence of MEcoRe commands and appendant outputs (marked by ▶) illustrates how the particular results from Example 1 to Example 5 have been determined:

// define a set of rules (cf. Example 1)
R := ((y|s)[0.8], (y|u)[0.7])

// start with an initial belief base consisting of the uniform distribution and $\mathcal{R}$,
// and compute the induced epistemic state (cf. Example 3)
currBBase := initBeliefBase(R);
▶ $P(syu) = 0.1950$, $P(sy\bar{u}) = 0.1758$, $P(s\bar{y}u) = 0.0408, \ldots$

// query the epistemic state induced by the current belief base
// about the conditional $(y|s \wedge u)$ (cf. Example 2)
currBBase.query((y|s ∧ u));
▶ $(y|s \wedge u)[0.8270]$

// revise the current belief base by $(s)[0.4]$
// and compute the induced epistemic state (cf. Example 4)
currBBase.revise((s)[0.4]);

▶ $P(syu) = 0.1703$, $P(s y \bar{u}) = 0.1496$, $P(s \bar{y} u) = 0.0340, \ldots$

// query the current induced epistemic state about $(y|u)$ (cf. Example 4)
currBBase.query((y|u));

▶ $(y|u)[0.7]$

// update the current belief base by $(y|s)[0.8]$
// and compute the induced epistemic state (cf. Example 5)
currBBase.update((y|s)[0.8]);

▶ $P(syu) = 0.1874$, $P(s y \bar{u}) = 0.1646$, $P(s \bar{y} u) = 0.0166, \ldots$

// query the current induced epistemic state about $(y|u)$ (cf. Example 5)
currBBase.query((y|u));

▶ $(y|u)[0.73]$

Hence, one is able to use both previously defined rule sets and rules that are entered just when they are needed, and combinations of both. The ability to manipulate rule sets, to automate sequences of updates and revisions, and to output selected (intermediate) results for comparing, yields a very expressive command language. This command language is a powerful tool for experimenting and testing with different setups. All core functions of the MEcoRe system are also accessible through a software interface (in terms of a Java API). So MEcoRe can easily be extended by a GUI or be integrated into another software application.

4.4 Related Work

There are many systems performing inferences in probabilistic networks, especially in Bayesian networks [51]. Some software systems supporting knowledge representation and reasoning with maximum entropy are SPIRIT [54], PIT [56], and the approach described in [55]. SPIRIT has been used in different applications involving credit analysis, investment business, or diagnosis in Chinese medicine, and LEXMED is a successfully working system for the diagnosis of appendicitis implemented in PIT. Graph based methods as they are used e.g. in SPIRIT [54] to implement reasoning at optimum entropy, are known to feature a very efficient representation of probability distributions via junction trees and hypergraphs, whereas the current version of MEcoRe works on a model based representation of probabilities. While this is less efficient, the primary aim of the MEcoRe project is to implement subjective probabilistic reasoning, as it could be performed by agents, making various belief operations possible. In particular, it allows changing of beliefs in a very flexible way by taking new, complex information into account. This is not possible with graph-based systems for probabilistic inference, as efficient methods of restructuring probabilistic networks as they would be required for the framework described in this paper still have to be developed.

Logical approaches to probabilistic reasoning have been made popular in artificial intelligence by Nilsson [44] and various probabilistic logics have been considered since then [22, 48, 3]. Halpern classified probabilistic logics in three types dependent on the way in which uncertainty is incorporated [19]. Whereas logics of the first type introduce probabilities over the elements of the domain of the logical language, logics of the second type consider probabilities over possible worlds. The third type combines the first and second type. Like many recent approaches [38, 11, 39], our formalism belongs to the second type. Reasoning under optimum entropy has been investigated extensively, see, e.g., [48, 30, 31]. Another popular approach is to consider all probability distributions that satisfy a knowledge base to derive probability intervals for queries [20, 37, 30]. Other successful approaches to probabilistic reasoning are based on graphical models [34, 32, 11] or extensions of classical probabilistic programming techniques [33].

5 Modelling and Reasoning with Probabilistic Logic in a Medical Application

In order to illustrate the use of the maximum entropy principle for probabilistic conditional logic and MECoRe's reasoning facilities, this section reports on a case study carried out in a medical scenario involving clinical patient data in neurosurgery. By modelling clinical brain tumor data and physicians' expert knowledge, we developed a prototype that could serve as a decision support system by giving informative feedback to the physicians who are responsible for deciding about the medical treatment of a patient. The evaluation of the results computed by MECoRe was carried out by a physician in neurosurgery by assessing MECoRe's output on a real documented patient case and deciding whether or not the system's results are plausible according to his medical knowledge and experience.

5.1 Brain Tumors

In this paper, we use the term brain tumor to refer to intracranial tumors which are tumorous neoplasms localized in the brain or its meningeal tissues. Two major clinical and neurophysiological problems are caused by a growing brain tumor process. One is the local infiltration of tumor tissue which destroys closely spaced brain tissue. Another one is caused by the increase of global intracranial pressure leading to a comprehensive brain damage. This is due to the fact that the cranium can be seen as a rigid box, since after birth the cranial fontanels start to ossify leaving the whole brain with very limited pressure releasing openings.

The prevalence of brain tumors is about 50:100.000 in the middle European region [52]. The incidence is about 1:10.000 per year. There are two age peaks, one within the range of 40 years and 70 years, another one within the childhood. Noteworthy, in childhood, brain tumors are the second most common tumor entity after leukemia. While adult patients mostly suffer from gliomas, meningeomas and metastases of other primary tumors, children mostly suffer from medulloblastomas, cerebellar astrozytomas and ependymomas [8]. Guiding symptoms are neurological failures; the brain tumor itself is mostly confirmed by medical imaging through CT/MRI-Scans. A histopathological tissue analysis secures the diagnosis and the exact classification-type and the grading of the brain tumor. Depending on the exact tumor type, the treatment consists of surgical remowement and/or chemotherapy. In rare cases of very small tumor sizes and probable benignity, surgery can be avoided if repeated medical imaging does not show any malignant potential within the next months.

5.2 Modelling Clinical Brain Tumor Data

For generating an initial knowledge base for clinical brain tumor data we will use various binary and multi-valued variables considering aspects of the patient, the patient's medical history, the observed symptoms, the possible diagnosis, etc; a medical justification for these variables and their values along with references to the relevant medical literature is given in [60, 61]. Since the prevalence of different tumor types varies with the age of patients, the variable `age` distinguishes patients with respect to the three values `le20` (less or equal 20 years old), `20to80` (between 20 and 80 years), and `ge80` (greater or equal 80 years). The binary variable `warningSymptoms` is true iff warning symptoms like perceptual disturbances or unusual pain in the head are present. Given results of a magnetic resonance tomography (MRT), the variable `malignancy` corresponds to the assumed malignancy of the tumor with respect to the WHO grading system [36]; a higher index corresponds to a higher malignancy. The binary variable `icpSymptoms` indicates whether MRT results provide symptoms for increased intracranial pressure (ICP). The preoperative physical fitness of patients is evaluated by the ASA (American Society of Anesthesiologists) classification system represented by the variable `ASA`. It is associated with perioperative risks, and a higher value indicates a higher risk. Only the first four states are considered here, as treatment of a brain tumor is of low priority for a higher value. Thus, so far we have:

```
              age : le20, 20to80, ge80
  warningSymptoms : true, false
       malignancy : 1, 2, 3, 4, other
      icpSymptoms : true, false
              ASA : 1, 2, 3, 4
```

In BT, the ten most common brain tumor types like gliomas and meningiomas [50] are taken into account. Together with the value other for any other tumor types, these brain tumor types constitute the values of the variable diagnosis:

```
        diagnosis :   glioblastoma,
                      pilocytic-astrocytoma,
                      diffuse-astrocytoma,
                      anaplastic-astrocytoma,
                      oligodendroglioma,
                      ependymoma,
                      meningeoma,
                      medulloblastoma,
                      cranialnerve-tumor,
                      metastatic-tumor,
                      other
```

Finally, there are three variables denoting the therapy, possible complications, and the expected health of the patient. The variable

```
       therapy : conservative, surgery, none
```

refers to the therapy to be chosen. We distinguish a conservative therapy without surgery, surgery, or no therapy at all. Possible complications during an inpatient stay are expressed by the variable

```
            complication : 1, 2, 3
```

which distinguishes the three stages 1 (no complications or minor, completely reversible complications like temporary pain after surgery), 2 (medium or heavy complications with uncertain reversibility like neurological or other functional disorders), and 3 (life-threatening complications like serious internal bleeding or neurological deficits at the risk of brain death). Thus, higher values correspond to more serious complications. The expected health of the patient after inpatient stay is denoted by:

diagnosis	Adults	Children
glioma		
- glioblastoma	15%	*unspecified*
- pilocytic-astrocytoma	*unspecified*	35%
- diffuse-astrocytoma	10%	*unspecified*
- anaplastic-astrocytoma	10%	*unspecified*
- oligodendroglioma	10%	*unspecified*
- ependymoma	4%	8%
meningeoma	20%	*unspecified*
medulloblastoma	7%	25%
cranialnerve-tumor	7%	*unspecified*
metastatic-tumor	10%	*unspecified*
other	*unspecified*	*unspecified*

Table 2: Empirical frequencies of brain tumor types, where *unspecified* stands for rare or unknown (collected from [8, 21, 43, 59]).

> prognosis : very_good, good, intermediate, poor, very_poor

The knowledge base BT uses these nine propositional variables as its vocabulary to represent clinical brain tumor data and corresponding expert knowledge. Note that although we have only 9 variables, due to the multiple values they induce $2^2 \times 3^3 \times 4 \times 5^2 \times 11 = 118.800$ possible worlds.

5.3 Initialization

There are various publications containing empirical frequencies of certain brain tumor types. For our initial version of our knowledge base BT, we encode the frequencies given in Table 2 that are collected from [8, 21, 43, 59] and that are given relative to the patient being an adult (age=20to80 or age=ge80) or being a child (age=le20). The representation of these frequencies is given by conditionals of the form

> (diagnosis=glioblastoma | !(age=le20))[0.15]
> (diagnosis=pilocytic-astrocytoma | age=le20)[0.35]
> $\vdots$

(diagnosis=cranialnerve-tumor | !(age=1e20))[0.07]

(diagnosis=metastatic-tumor | !(age=1e20))[0.10]

where, using the input syntax of MEcoRe, ! denotes negation. Additionally, BT contains the probabilistic facts

(age=1e20)[0.15]

(age=20to80)[0.62]

reflecting the age distribution in Germany in the year 2009.

Note that there are some missing frequencies in Table 2, and thus, there are no conditionals in BT for these missing frequencies. In order to obtain a full probability distribution over all variables and their values, the missing knowledge is completed in an information-theoretically optimal way by employing the principle of maximum entropy, thus by being as unbiased as possible with respect to each diagnosis with unspecified probability. In MEcoRe, the computation of the epistemic state induced by the belief base which incorporates the knowledge given by BT is started by

line 1:   currBBase := initBeliefBase(BT);

so that the belief base currBBase induces the maximum entropy distribution over BT.

In order to be able to ask a set of queries instead of just a single query at the same time, MEcoRe allows the introduction of an identifier to denote a set of queries. Here, we will illustrate this feature with a singleton set containing an unquantified conditional for the diagnosis under the premise that the patient is older than 80 and that he suffers from warning symptoms

line 2:   queriesBT := (diagnosis|(age=ge80) ∧ warningSymptoms);
line 3:   currBBase.query(queriesBT);

which yields the following probabilities:

diagnosis	probability	diagnosis	probability
glioblastoma	0.150	meningeoma	0.200
pilocytic-astrocytoma	0.035	medulloblastoma	0.070
diffuse-astrocytoma	0.100	cranialnerve-tumor	0.070
anaplastic-astrocytoma	0.100	metastatic-tumor	0.100
oligodendroglioma	0.100	other	0.035
ependymoma	0.040		

Note that up to now, BT does not contain any information about the influence of warning symptoms or the observation that the patient is more than 80 years old. Therefore, in the maximum entropy distribution induced by `currBBase`, the corresponding premise given in the queries in `queriesBT` (cf. line 2) does not cause a deviation from the probabilities given in the original conditionals in BT and taken from Table 2. Note also that the prababilities for the two possible diagnosis values `pilocytic-astrocytoma` and `other` missing for adults in Table 2 have also been computed as expected.

5.4 Revising BT

Besides available statistical data, another important knowledge source is the clinical expert knowledge of a physician. For example, for adults, Table 2 tells us that the most frequently appearing glioma tumor type is `glioblastoma`, but no information is provided about its probability given specific symptoms. An experienced physician working with brain tumor patients might state the following conditionals expressing his expert beliefs about the probability of a `glioblastoma` given various observations:

$$(\text{diagnosis=glioblastoma} \mid !(\text{age=1e20}) \land \text{warningSymptoms})[0.20] \quad (16)$$
$$(\text{diagnosis=glioblastoma} \mid !(\text{age=1e20}) \land \text{icpSymptoms})[0.20] \quad (17)$$
$$(\text{diagnosis=glioblastoma} \mid !(\text{age=1e20}) \land (\text{malignancy=4}))[0.40] \quad (18)$$
$$(\text{diagnosis=glioblastoma} \mid !(\text{age=1e20}) \land (\text{malignancy=3}))[0.10] \quad (19)$$
$$(\text{diagnosis=glioblastoma} \mid !(\text{age=1e20}) \land (\text{malignancy=2}))[0.05] \quad (20)$$
$$(\text{diagnosis=glioblastoma} \mid !(\text{age=1e20}) \land (\text{malignancy=1})[0.01] \quad (21)$$

Taking into account only Table 2, the probability for `glioblastoma` is 15%. Therefore, given the respective preconditions, rules (16) - (18) would increase the probability, whereas rules (19) - (21) would decrease it.

In [60], about 90 conditionals expressing such expert knowledge from a physician's point of view are formulated. With `expertBT` denoting the set of these conditionals, we will incorporate this new knowledge into the current belief base. We can achieve this in such a way as if it had been available already in the original knowledge base BT by using the belief change operation *revision* (cf. Section 3 and [29]). In MECORE, this is easily expressed by

line 4: `currBBase.revise(expertBT);`

Now, asking the `queriesBT` (cf. line 2) again, the probabilities have changed considerably in the new epistemic state induced by the revised belief base:

diagnosis	probability	diagnosis	probability
glioblastoma	0.223	meningeoma	0.156
pilocytic-astrocytoma	0.050	medulloblastoma	0.065
diffuse-astrocytoma	0.098	cranialnerve-tumor	0.057
anaplastic-astrocytoma	0.106	metastatic-tumor	0.106
oligodendroglioma	0.086	other	0.011
ependymoma	0.039		

For example, the probability for **glioblastoma** increased from 15% to 22.3%, while the probability for **meningeoma** decreased from 20% to 15.6%. This is well in accordance with the observations made by physicians working in this area [60].

For illustrating MECoRe's copying mechanism for belief bases, let us assume that we want to store the current belief base and its associated epistemic state for later use. This is achieved by instructing MECoRe to make a copy:

line 5:   revBB := currBBase.copy();

In this way, we can perform any further change operations on **currBBase** and can still come back to the belief base obtained so far by just using **revBB**.

5.5 Updating BT

Now suppose that later on, there has been some change in the environment and experts think that the conditionals (16) - (18) hold with other probabilities. So let **gliobNew** denote the set consisting of the following conditionals, which have modified probabilities:

(diagnosis=glioblastoma | !(age=le20) ∧ warningSymptoms)[0.15]   (22)

(diagnosis=glioblastoma | !(age=le20) ∧ icpSymptoms)[0.25]   (23)

(diagnosis=glioblastoma | !(age=le20) ∧ (malignancy=4))[0.45]   (24)

Revision of the current belief base with **gliobNew** would lead to an inconsistency since the conditionals (16) - (18) and the conditionals (22) - (24) in **gliobNew** cannot be satisfied simultaneously. However, MECoRe's update operation of **currBBase** by **gliobNew** can incorporate the new knowledge into the current belief base. Afterwards, the epistemic state induced by the updated belief base satisfies **gliobNew** and has minimum cross entropy with respect to previous epistemic state (cf. Section 3). Note that update (in contrast to revision) is the more appropriate operation here, since the shift of the probabilities reflects a changed environment. So we perform the update operation in MECoRe by stating:

line 6:   currBBase.update(gliobNew);

Next, we could, e.g., address further queries to the updated belief base to analyze the effects of the update operation.

5.6 Prognosis and What-If-Analysis

Now, we come back to the belief base `revBB` obtained in Section 5.4, which has not been affected by the update operation in Section 5.5. For the real documented case [60] of a patient being older than 80 years, with `warningSymptoms`, `icpSymptoms`, and `malignancy=4`, we ask MEcoRe about the diagnosis:

line 7:   revBB.query((diagnosis | (age=ge80) ∧ warningSymptoms
                              ∧ icpSymptoms ∧ (malignancy=4)));

MEcoRe returns a probability of 55.6% for the diagnosis glioblastoma, being very plausible from a physician's point of view. Assuming that glioblastoma were indeed the correct diagnosis and assuming further that a surgery would be chosen, the prognosis for complications that might occur are determined by:

line 8:   hypothesis := ((diagnosis=glioblastoma)[1.0],
                              (therapy=surgery)[1.0]);
line 9:   whatIfQ := (complication | (age=ge80) ∧ warningSymptoms
                              ∧ icpSymptoms ∧ (malignancy=4));
line 10: revBB.whatif(hypothesis, whatIfQ);

Note that what-if is similar to an update except that it does not change the current belief base. The resulting probabilities for complications of grade 1, 2, and 3 are 0.4%, 45.4%, and 54.2%, respectively. While complications of grade 2 or 3 are rare in general, the provided evidence and the given assumptions caused MEcoRe to rise the probabilities for these types of complications considerably. After surgical treatment of the given patient, there was indeed a complication of grade 2. From a clinical perspective, the probabilities for `complication` computed by MEcoRe is an adequate warning; however, the probability for grade 3 is a bit too pessimistic, since compared to similar patient-risk constellations, life-threatening complications are frequent, but less than 50%. Here, a corresponding adaptation of the conditionals constraining the probabilities for grade 3 complications might lead to a more realistic probability value for this query. Further types of queries for BT asking MEcoRe for the expected health of patients after inpatient stay, returned a very realistic prognosis from a medical point of view [60].

6 Conclusions and Further Work

In this paper, we proposed a framework for powerful knowledge and belief management operations as needed by an intelligent agent acting in an uncertain, evolving environment. Using probabilistic logic with optimum entropy semantics, the core functionality of a change operator based on minimum cross entropy supports revision, update, and iterative belief changes in a setting where epistemic states are modified by sets of conditionals, thus exceeding by far the classical AGM theory and related approaches dealing only with propositional beliefs. We gave an overview on the MEcoRe system realizing and implementing the agent model, including diagnosis and hypothetical reasoning. As an application scenario for MEcoRe, we reported on a case study using probabilistic conditional logic and the principle of maximum entropy to model clinical brain tumor data and medical expert knowledge in neurosurgery. The resulting knowledge base contains approximately 110 probabilistic conditionals over nine multi-valued variables that medical experts identified to be at the core of clinical brain tumor data analysis. Using MEcoRe for working with this knowledge base produced realistic probabilities for diagnosis and prognosis from a clinical physician's point of view.

In future work, we plan to further investigate and elaborate the formal properties of the knowledge and belief management operations, to employ graph-based methods for representing probability distributions in MEcoRe, and to extend and refine the medical modelling in the brain tumor application. For representing uncertainty or ignorance about probabilities, probability intervals could be used. Thus, a further area of our future work is to investigate whether it is possible to extend the framework presented in this article by considering not only point probabilities, but allowing for stating lower and upper probabilities as well.

References

[1] E. Adams. *The Logic of Conditionals*. D. Reidel, Dordrecht, 1975.

[2] C. Alchourrón, P. Gärdenfors, and D. Makinson. On the logic of theory change: Partial meet contraction and revision functions. *Journal of Symbolic Logic*, 50(2):510–530, 1985.

[3] F. Bacchus, A. Grove, J. Halpern, and D. Koller. From statistical knowledge bases to degrees of belief. *Artificial Intelligence*, 87(1-2):75–143, 1996.

[4] C. Beierle, M. Finthammer, N. Potyka, J. Varghese, and G. Kern-Isberner. A case study on the application of probabilistic conditional modelling and reasoning to clinical patient data in neurosurgery. In L. C. van der Gaag, editor, *Symbolic and Quantitative Approaches to Reasoning with Uncertainty (ECSQARU 2013)*, volume 7958 of *LNCS*, pages 49–60. Springer, 2013.

[5] C. Beierle and G. Kern-Isberner. A conceptual agent model based on a uniform approach to various belief operations. In B. Mertsching, M. Hund, and Z. Aziz, editors, *Advances in Artificial Intelligence. 32nd Annual German Conference on AI (KI 2009)*, volume 5803 of *LNAI*, pages 273–280. Springer, 2009.

[6] S. Benferhat, D. Dubois, and H. Prade. Nonmonotonic reasoning, conditional objects and possibility theory. *Artificial Intelligence*, 92:259–276, 1997.

[7] S. Benferhat, D. Dubois, and H. Prade. Possibilistic and standard probabilistic semantics of conditional knowledge bases. *Journal of Logic and Computation*, 9(6):873–895, 1999.

[8] H. Bruch and O. Trentz. *Berchthold Chirurgie, 6.Auflage*. Elsevier GmbH, 2008.

[9] I. Csiszár. I-divergence geometry of probability distributions and minimization problems. *Ann. Prob.*, 3:146–158, 1975.

[10] A. Darwiche and J. Pearl. On the logic of iterated belief revision. *Artificial Intelligence*, 89:1–29, 1997.

[11] P. Domingos and D. Lowd. *Markov Logic: An Interface Layer for Artificial Intelligence*. Synthesis Lectures on Artificial Intelligence and Machine Learning. Morgan and Claypool, San Rafael, CA, 2009.

[12] D. Dubois and H. Prade. Conditional objects as nonmonotonic consequence relationships. *IEEE Transactions on Systems, Man and Cybernetics*, 24(12):1724–1740, 1994.

[13] D. Dubois and H. Prade. Focusing vs. belief revision: A fundamental distinction when dealing with generic knowledge. In *Proc. ECSQARU-FAPR'97*, pages 96–107, Berlin Heidelberg New York, 1997. Springer.

[14] R. Fagin and J. Y. Halpern. Reasoning about knowledge and probability. *J. ACM*, 41(2):340–367, 1994.

[15] M. Finthammer and C. Beierle. Using equivalences of worlds for aggregation semantics of relational conditionals. In B. Glimm and A. Krüger, editors, *KI 2012: Advances in Artificial Intelligence, 35th Annual German Conference on AI, Saarbrücken, Germany, September 24-27, 2012. Proceedings*, volume 7526 of *LNAI*, pages 49–60. Springer, 2012.

[16] V. Fischer and M. Schramm. Tabl – a tool for efficient compilation of probabilistic constraints. Technical Report TUM-19636, Technische Universität München, 1996.

[17] P. Gärdenfors. *Knowledge in Flux: Modeling the Dynamics of Epistemic States*. MIT Press, Cambridge, Mass., 1988.

[18] J. Halpern. *Reasoning About Uncertainty*. MIT Press, 2005.

[19] J. Y. Halpern. An analysis of first-order logics of probability. *Artificial Intelligence*, 46:311–350, 1990.

[20] P. Hansen and B. Jaumard. Probabilistic satisfiability. In J. Kohlas and S. Moral, editors, *Handbook of Defeasible Reasoning and Uncertainty Management Systems*, volume 5 of *Handbook of Defeasible Reasoning and Uncertainty Management Systems*, pages 321–367. Springer Netherlands, 2000.

[21] N. Hosten and T. Liebig. *Computertomografie von Kopf und Wirbelsäule*. Georg Thieme Verlag, 2007.

[22] B. Jaumard, P. Hansen, and M. Poggi. Column generation methods for probabilistic logic. *ORSA - Journal on Computing*, 3(2):135–148, 1991.

[23] E. Jaynes. Where do we stand on maximum entropy? In *Papers on Probability, Statistics and Statistical Physics*, pages 210–314. D. Reidel Publishing Company, Dordrecht, Holland, 1983.

[24] H. Katsuno and A. Mendelzon. On the difference between updating a knowledge base and revising it. In *Proceedings Second International Conference on Principles of Knowledge Representation and Reasoning, KR'91*, pages 387–394, San Mateo, Ca., 1991. Morgan Kaufmann.

[25] G. Kern-Isberner. Characterizing the principle of minimum cross-entropy within a conditional-logical framework. *Artificial Intelligence*, 98:169–208, 1998.

[26] G. Kern-Isberner. A note on conditional logics and entropy. *International Journal of Approximate Reasoning*, 19:231–246, 1998.

[27] G. Kern-Isberner. *Conditionals in nonmonotonic reasoning and belief revision*. Springer, Lecture Notes in Artificial Intelligence LNAI 2087, 2001.

[28] G. Kern-Isberner. A thorough axiomatization of a principle of conditional preservation in belief revision. *Annals of Mathematics and Artificial Intelligence*, 40(1-2):127–164, 2004.

[29] G. Kern-Isberner. Linking iterated belief change operations to nonmonotonic reasoning. In G. Brewka and J. Lang, editors, *Proceedings 11th International Conference on Knowledge Representation and Reasoning, KR'2008*, pages 166–176, Menlo Park, CA, 2008. AAAI Press.

[30] G. Kern-Isberner and T. Lukasiewicz. Combining probabilistic logic programming with the power of maximum entropy. *Artificial Intelligence, Special Issue on Nonmonotonic Reasoning*, 157(1-2):139–202, 2004.

[31] G. Kern-Isberner and M. Thimm. Novel semantical approaches to relational probabilistic conditionals. In F. Lin, U. Sattler, and M. Truszczynski, editors, *Proceedings Twelfth International Conference on the Principles of Knowledge Representation and Reasoning, KR'2010*, pages 382–391. AAAI Press, 2010.

[32] K. Kersting and L. De Raedt. Bayesian Logic Programming: Theory and Tool. In L. Getoor and B. Taskar, editors, *An Introduction to Statistical Relational Learning*. MIT Press, 2007.

[33] A. Kimmig, B. Demoen, L. De Raedt, V. S. Costa, and R. Rocha. On the implementation of the probabilistic logic programming language problog. *Theory and Practice of Logic Programming*, 11(2-3):235–262, 2011.

[34] D. Koller and N. Friedman. *Probabilistic Graphical Models: Principles and Techniques*. MIT Press, 2009.

[35] D. C. Liu and J. Nocedal. On the limited memory BFGS method for large scale optimization. *Mathematical Programming*, 45:503–528, 1989.

[36] D. N. Louis, H. Ohgaki, O. D. Wiestler, W. K. Cavenee, P. C. Burger, A. Jouvet, B. W. Scheithauer, and P. Kleihues. The 2007 WHO Classification of Tumours of the Central

Nervous System. *Acta Neuropathologica*, 114(2):97–109, 2007.

[37] T. Lukasiewicz. Probabilistic deduction with conditional constraints over basic events. *Journal of Artificial Intelligence Research*, 10:380–391, 1999.

[38] T. Lukasiewicz. Expressive probabilistic description logics. *Artificial Intelligence*, 172(6):852–883, 2008.

[39] C. Lutz and L. Schröder. Probabilistic description logics for subjective uncertainty. In *Proc. KR 2010*. AAAI Press, 2010.

[40] R. Malouf. A comparison of algorithms for maximum entropy parameter estimation. In *proceedings of the 6th conference on Natural language learning - Volume 20*, COLING-02, pages 1–7, Stroudsburg, PA, USA, 2002. Association for Computational Linguistics.

[41] C.-H. Meyer. *Korrektes Schließen bei unvollständiger Information*. Peter Lang Verlag, 1998.

[42] T. P. Minka. A comparison of numerical optimizers for logistic regression. Technical Report, Microsoft Research, 2003 (rev. 2007).

[43] M. Müller. *Chirurgie für Studium und Praxis, 9. Auflage*. Medizinische Verlags- und Informationsdienste, 2007.

[44] N. Nilsson. Probabilistic logic. *Artificial Intelligence*, 28:71–87, 1986.

[45] D. Nute. *Topics in Conditional Logic*. D. Reidel Publishing Company, Dordrecht, Holland, 1980.

[46] J. Paris. *The uncertain reasoner's companion – A mathematical perspective*. Cambridge University Press, 1994.

[47] J. Paris. Common sense and maximum entropy. *Synthese*, 117:75–93, 1999.

[48] J. Paris and A. Vencovská. A note on the inevitability of maximum entropy. *International Journal of Approximate Reasoning*, 14:183–223, 1990.

[49] J. Paris and A. Vencovska. In defence of the maximum entropy inference process. *International Journal of Approximate Reasoning*, 17(1):77–103, 1997.

[50] B. J. Park, H. K. Kim, B. Sade, and J. H. Lee. Epidemiology. In J. H. Lee, editor, *Meningiomas: Diagnosis, Treatment, and Outcome.*, page 11. Springer, 2009.

[51] J. Pearl. *Probabilistic Reasoning in Intelligent Systems*. Morgan Kaufmann, San Mateo, Ca., 1988.

[52] K. Poeck and W. Hacke. *Neurologie*. Springer DE, 2006.

[53] N. Potyka, D. Marenke, and E. Mittermeier. An overview of algorithmic approaches to compute optimum entropy distributions in the expert system shell mecore. In *4th Workshop on Dynamics of Knowledge and Belief (DKB-2013)*, pages 61–74, 2013.

[54] W. Rödder, E. Reucher, and F. Kulmann. Features of the expert-system-shell SPIRIT. *Logic Journal of the IGPL*, 14(3):483–500, 2006.

[55] F. Schmidt, J. Gebhardt, and R. Kruse. Handling revision inconsistencies: Towards better explanations. In S. Destercke and T. Denoeux, editors, *Symbolic and Quantitative Approaches to Reasoning with Uncertainty - 13th European Conference, ECSQARU 2015, Compiègne, France, July 15-17, 2015. Proceedings*, volume 9161 of *LNCS*, pages

257–266. Springer, 2015.

[56] M. Schramm and W. Ertel. Reasoning with probabilities and maximum entropy: the system PIT and its application in LEXMED. In *Symposium on Operations Research, SOR'99*, 1999.

[57] J. Shore. Relative entropy, probabilistic inference and AI. In L. Kanal and J. Lemmer, editors, *Uncertainty in Artificial Intelligence*, pages 211–215. North-Holland, Amsterdam, 1986.

[58] J. Shore and R. Johnson. Axiomatic derivation of the principle of maximum entropy and the principle of minimum cross-entropy. *IEEE Transactions on Information Theory*, IT-26:26–37, 1980.

[59] H.-J. Steiger and R. H.J. *Manual Neurochirurgie*. Ecomed Medizin, 2006.

[60] J. Varghese. Using probabilistic logic for the analysis and evaluation of clinical patient data in neurosurgery. B.Sc. Thesis, Univ. Hagen, 2012. (in German).

[61] J. Varghese, C. Beierle, N. Potyka, and G. Kern-Isberner. Using probabilistic logic and the principle of maximum entropy for the analysis of clinical brain tumor data. In *Proc. 26th IEEE International Symposium on Computer-Based Medical Systems (CBMS 2013)*, pages 401–404. IEEE Press, New York, 2013.

www.ingramcontent.com/pod-product-compliance
Lightning Source LLC
Chambersburg PA
CBHW080437110426
42743CB00016B/3193